Transactions on Computer Systems and Networks

Transactions on Computer Systems and Networks is a unique series that aims to capture advances in evolution of computer hardware and software systems and progress in computer networks. Computing Systems in present world span from miniature IoT nodes and embedded computing systems to large-scale cloud infrastructures, which necessitates developing systems architecture, storage infrastructure and process management to work at various scales. Present day networking technologies provide pervasive global coverage on a scale and enable multitude of transformative technologies. The new landscape of computing comprises of self-aware autonomous systems, which are built upon a software-hardware collaborative framework. These systems are designed to execute critical and non-critical tasks involving a variety of processing resources like multi-core CPUs, reconfigurable hardware, GPUs and TPUs which are managed through virtualisation, real-time process management and fault-tolerance. While AI, Machine Learning and Deep Learning tasks are predominantly increasing in the application space the computing system research aim towards efficient means of data processing, memory management, real-time task scheduling, scalable, secured and energy aware computing. The paradigm of computer networks also extends it support to this evolving application scenario through various advanced protocols, architectures and services. This series aims to present leading works on advances in theory, design, behaviour and applications in computing systems and networks. The Series accepts research monographs, introductory and advanced textbooks, professional books, reference works, and select conference proceedings.

Balvinder Shukla · K. M. Soni · Nitasha Hasteer
Editors

Sustainable Technologies for Transforming Businesses and Societies

The Keynote Lectures of the 6th International Conference on Entrepreneurship, Innovation and Leadership (ICEIL 2024)

Springer

Editors
Balvinder Shukla
Amity University Uttar Pradesh
Noida, Uttar Pradesh, India

K. M. Soni
Amity University Uttar Pradesh
Noida, Uttar Pradesh, India

Nitasha Hasteer
Amity University Uttar Pradesh
Noida, Uttar Pradesh, India

ISSN 2730-7484 ISSN 2730-7492 (electronic)
Transactions on Computer Systems and Networks
ISBN 978-981-96-9470-9 ISBN 978-981-96-9471-6 (eBook)
https://doi.org/10.1007/978-981-96-9471-6

This Springer imprint is published by the registered company Springer Nature Singapore Pte Ltd.
The registered company address is: 152 Beach Road, #21-01/04 Gateway East, Singapore 189721, Singapore

Preface

In the current fast-changing technological environment, countries have unique chances to attain self-sufficiency by thoughtfully integrating and utilizing advanced innovations. Disruptive technologies in engineering domain such as artificial intelligence, blockchain, the Internet of Things, bioinformatics, gene editing, tissue engineering, and 3D printing are transforming industries and communities around the globe. In conjunction with these advancements, fostering and supporting indigenous technologies is vital for stimulating economic progress. By capitalizing on local innovations designed to meet regional demands, countries can develop robust economies, reduce reliance on imports, and formulate scalable, sustainable responses to key challenges.

A self-reliant nation focuses on domestic capacity building, technology development, and export competitiveness and lays the groundwork for realizing economic growth ambitions. Indigenous innovations in engineering and technology and good governance are key factors that make a nation self-reliant. Researchers, scientists, practitioners, and policymakers across the globe are leveraging transformative technologies like artificial intelligence, machine learning, and blockchain, to develop digital solutions and frameworks for the future. This book is a compilation of keynote talks delivered by esteemed experts, showcasing their significant contributions to developing intelligent IT solutions that drive economic growth, foster innovation, and address pressing environmental and societal challenges. It offers timely insights and knowledge for those striving to advance in engineering and contribute to building robust, self-sustaining economies. The content of this volume is drawn from the keynote talks presented during the 6th edition of the International Conference on Entrepreneurship, Innovation, and Leadership (ICEIL 2024). The topics covered in this volume reflect a harmonious blend of technology, economy, and societal impact, demonstrating the interconnected nature of modern innovation. Beginning with Dyadic Human-Robot Interaction: Opportunities and Challenges, the talks explore the transformative potential of robotics in human-centric applications. This is complemented by approaching Metaheuristics from a Bibliometric Point of View, which delves into optimization techniques shaping decision-making and resource management across domains. The discussion transitions to societal implications with Queuing for Civility: Regulating Emotions and Reducing Toxicity in Digital Discourse, offering insights into fostering healthier digital interactions. On the technological front, talks such as Effect of Multiwalled Carbon Nanotubes on the Compressive Strength and Microstructural Properties of Rubberized Concrete Using RSM Modelling highlight advancements in material science and sustainable construction. Meanwhile, a talk on complex fuzzy graphs and its applications presents novel mathematical frameworks with far-reaching applications. The compilation concludes with South African Perceptions on Artificial Intelligence in the Context of Morality and Sentience, offering a regional perspective on the ethical and economic implications of AI in modern society. This volume is not just a collection of ideas but a cohesive narrative showcasing how technology can transform economies and societies. It aims to inspire graduate students, researchers, scientists, practitioners, and policymakers to explore innovative solutions and contribute meaningfully to sustainable development and economic resilience.

We thank the management of Amity University, Uttar Pradesh, who provided us the opportunity to publish the book. We would also like to extend our gratitude to the expert reviewers who have helped us as part of our technical programme committee. We wish to acknowledge and thank the worthy speakers from across the globe who delivered keynote talks, contributed their chapters and cooperated with us at every stage of compilation and publication. We owe our sincere gratitude to all our family members and friends who helped us through this journey of publishing a book. Our appreciation goes to every individual who supported us in this endeavour. Finally, we are grateful to the editing team of Springer who provided immense support to us in the compilation of the book and shaped it into a marketable product.

Noida, India

Dr. Balvinder Shukla
Dr. K. M. Soni
Dr. Nitasha Hasteer

Contents

Contributors

Samuel Adebayo Centre for Intelligent Autonomous Manufacturing Systems (i-AMS), Queen's University Belfast, Belfast, Northern Ireland, UK;
School of Electronics, Electrical Engineering and Computer Science, Queen's University Belfast, Belfast, Northern Ireland, UK

Mehmet Anıl Akbay Artificial Intelligence Research Institute (IIIA-CSIC), Bellaterra, Barcelona, Spain

Adnan Anwar School of Information Technology, Deakin University, Geelong, VIC, Australia

Vasudevan Arveendh Department of Civil and Environmental Engineering, Universiti Teknologi PETRONAS, Bandar Seri Iskandar, Perak, Malaysia

Abhijeet Vidyadhar Baikerikar Department of Civil and Environmental Engineering, Universiti Teknologi PETRONAS, Bandar Seri Iskandar, Perak, Malaysia

Assia Bakali Ecole Royale Navale, Casablanca, Morocco

Naraindas Bheel Department of Civil and Environmental Engineering, Universiti Teknologi PETRONAS, Bandar Seri Iskandar, Perak, Malaysia

Christian Blum Artificial Intelligence Research Institute (IIIA-CSIC), Bellaterra, Barcelona, Spain

Camilo Chacón Sartori Artificial Intelligence Research Institute (IIIA-CSIC), Bellaterra, Barcelona, Spain

Joost C. Dessing Centre for Intelligent Autonomous Manufacturing Systems (i-AMS), Queen's University Belfast, Belfast, Northern Ireland, UK;
School of Psychology, Queen's University Belfast, Belfast, Northern Ireland, UK

Shama Islam School of Engineering, Deakin University, Geelong, VIC, Australia

Seán McLoone Centre for Intelligent Autonomous Manufacturing Systems (i-AMS), Queen's University Belfast, Belfast, Northern Ireland, UK;
School of Electronics, Electrical Engineering and Computer Science, Queen's University Belfast, Belfast, Northern Ireland, UK

Valeh Moghaddam School of Information Technology, Deakin University, Geelong, VIC, Australia

Bashar S. Mohammed Department of Civil and Environmental Engineering, Universiti Teknologi PETRONAS, Bandar Seri Iskandar, Perak, Malaysia

S. Mohanaselvi Department of Mathematics, SRM Institute of Science and Technology, Kattankulathur, India

Matthew Reid Center for IT and National Development in Africa, University of Cape Town, Rondebosch, South Africa

Jaume Reixach Artificial Intelligence Research Institute (IIIA-CSIC), Bellaterra, Barcelona, Spain

Guillem Rodríguez Corominas Artificial Intelligence Research Institute (IIIA-CSIC), Bellaterra, Barcelona, Spain

Broumi Said Laboratory of Information Processing, Faculty of Science Ben M'Sik, University of Hassan II, Casablanca, Morocco;
STIE Team Regional Center for the Professions of Education and Training (C.R.M.E.F), Casablanca, Settat, Morocco

Mohamed Talea Laboratory of Information Processing, Faculty of Science Ben M'Sik, University of Hassan II, Casablanca, Morocco

Jean-Paul Van Belle Center for IT and National Development in Africa, University of Cape Town, Rondebosch, South Africa

Akriti Verma School of Engineering, Deakin University, Geelong, VIC, Australia

Chapter 1
Dyadic Human-Robot Interaction: Emerging Technologies, Challenges, and Opportunities

Samuel Adebayo, Joost C. Dessing, and Seán McLoone

Abstract This paper explores the current state and future prospects of dyadic Human-Robot Interaction (HRI), focusing on the opportunities and challenges in this evolving field. We examine the role of key technologies, including computer vision, Digital Twins (DTs), and Generative Artificial Intelligence (GenAI), in enhancing the quality and effectiveness of dyadic interactions between humans and robots. Beginning with a contextualisation of dyadic HRI and its foundational concepts, we identify and review the stages of dyadic interaction and how they are all interconnected with the latest advancements in visual perception, and non-visual systems, then highlighting how these technologies can be employed to create more responsive and adaptable robots. Central to our discussion is the integration of DTs and GenAI and the exploration of how they can be leveraged to develop predictive models that improve real-time interactions. We also address the ethical, societal, and regulatory implications of these technological integrations, considering how they might shape the future of human-robot partnerships. Finally, we propose future research directions aimed at advancing the field of dyadic HRI. By providing this analytical review, this paper seeks to offer a concise roadmap for the future of dyadic HRI where technology and human-centred design converge.

S. Adebayo · J. C. Dessing · S. McLoone (✉)
Centre for Intelligent Autonomous Manufacturing Systems (i-AMS), Queen's University Belfast, Belfast, Northern Ireland, UK
e-mail: s.mcloone@qub.ac.uk

S. Adebayo
e-mail: sadebayo01@qub.ac.uk

J. C. Dessing
e-mail: j.dessing@qub.ac.uk

S. Adebayo · S. McLoone
School of Electronics, Electrical Engineering and Computer Science, Queen's University Belfast, Belfast, Northern Ireland, UK

J. C. Dessing
School of Psychology, Queen's University Belfast, Belfast, Northern Ireland, UK

B. Shukla et al. (eds.), *Sustainable Technologies for Transforming Businesses and Societies*, Transactions on Computer Systems and Networks,
https://doi.org/10.1007/978-981-96-9471-6_1

Keywords Digital twin · Human-robot interaction · Generative AI · Visual cues · Dyadic interaction · Computer vision · Machine learning

1.1 Introduction

Human-robot interaction (HRI) is a rapidly growing field at the intersection of robotics, artificial intelligence, and cognitive science. As robots become increasingly integrated into sectors such as manufacturing, healthcare, education, and assisted living, the need for advanced and effective human-robot collaboration has intensified. The development of human-centric design in robotics is poised to drive the next industrial revolution, Industry 5.0, an evolution of Industry 4.0 that refocuses the adoption of advanced technologies, including AI and robotics, towards achieving human-centricity, sustainability, and resilience in manufacturing systems (Leng et al. 2022; Nahavandi 2019; Industry 5.0—European Commission 2024; Xu et al. 2021).

At the heart of this evolution is dyadic interaction—a core aspect of HRI that emphasises two-way, reciprocal communication between a human and a robot. This interaction is essential for human-robot collaboration, whether in a group setting or one-on-one. Effective communication and collaboration between humans and robots depend on this dyadic nature, ensuring that robots can understand and respond to human behaviours, intentions, and contextual cues in a meaningful and adaptive way.

As illustrated in Fig. 1.1, the key technologies involved in dyadic interaction between humans and robots, and by extension, Industry 5.0, include Computer Vision, NLP, Proprioception, Digital Twins, Multimodal Integration, and Artificial Intelligence. These technologies enable robots to understand and respond to the dynamic and stochastic nature of our world. Specifically, integrating multimodal systems (encompassing computer vision, speech recognition, tactile feedback, and AI) into robotic frameworks are essential for enhancing the quality of interactions. Such systems allow robots to perceive and interpret a diverse range of human cues, including body language, facial expressions, speech patterns, and touch, facilitating more natural and intuitive exchanges. However, implementing these multimodal AI systems can be computationally and resource-intensive, making them impractical for real-time deployment. DTs offer a potential solution by creating real-time, virtual replicas of physical environments and human behaviours that enhance and accelerate AI learning and inform decision-making, enabling robots to adapt more effectively to dynamic and unpredictable scenarios.

As we approach Industry 5.0, integrating these advanced technologies will mark a significant turning point in the evolution of HRI. Their combined potential will enable robots to engage in increasingly complex dyadic interactions with humans. However, despite rapid progress, challenges persist in making these interactions seamless and contextually adaptive to the subtleties of human activities. This paper reviews the current state of these technologies, identifies ongoing challenges, and outlines future directions for advancing dyadic HRI as we transition into the Industry 5.0 era.

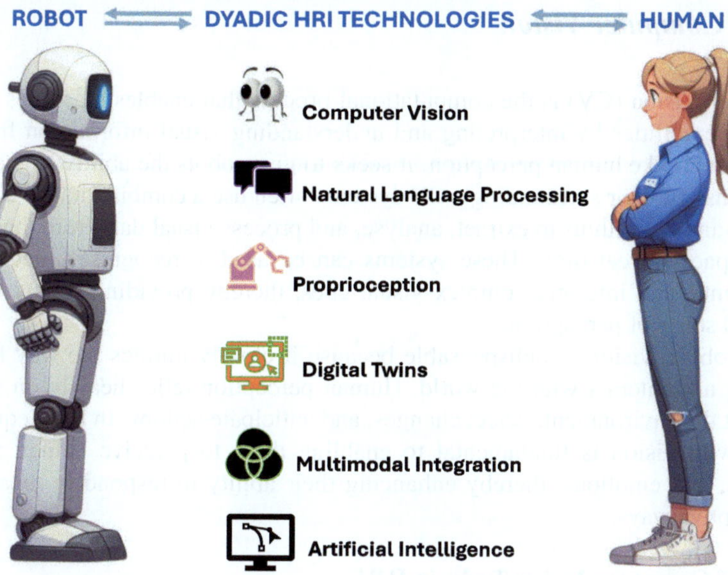

Fig. 1.1 Dyadic HRI technologies overview: showing the core technologies contributing to dyadic interaction between a robot and a human, including computer vision, natural language processing, proprioception, digital twins, multimodal integration, and artificial intelligence

1.2 Foundational Concepts in Dyadic HRI

Dyadic interaction involves two parties engaging in dynamic and reciprocal communication. In the context of HRI, this often means direct engagement between a human and a robot, where each party responds to the other's actions and cues. Whether through visual signals, speech, or touch, dyadic interaction forms the foundational basis for effective communication and collaboration in HRI, similar to interactions between two humans.

Without a doubt, dyadic interaction is crucial for designing robots that can engage in meaningful, responsive exchanges with humans. In HRI, the dyadic model emphasises mutual understanding, allowing robots to be both reactive and proactive, anticipating human actions, intentions, and contextual information. This fluid exchange of information, similar to natural human-to-human communication, fosters seamless collaboration and makes interactions intuitive.

This section explores the foundational technologies enabling dyadic interaction in HRI, including computer vision, speech recognition, tactile feedback, proprioception, and AI. We will also examine how Digital Twins (DTs) can enhance real-time interaction and adaptation, ensuring that robots can dynamically respond to changing human behaviours and environmental contexts.

1.2.1 Computer Vision

Computer Vision (CV) is the computational process that enables machines to gain perceptive abilities by interpreting and understanding visual information from the world. Much like human perception, it seeks to give robots the ability to "see" and make sense of their surroundings. CV systems often use a combination of cameras, sensors, and algorithms to extract, analyse, and process visual data from their interaction space in real time. These systems can be used to recognise objects, track movements, and interpret complex visual cues, thereby providing robots with an essential sense of perception.

For robots, vision is indispensable because it closely mimics the way humans perceive and interact with the world. Human perception relies heavily on sight to interpret the environment, detect changes, and anticipate actions. In HRI, equipping robots with vision is fundamental to enabling them to perceive human actions, gestures, and emotions, thereby enhancing their ability to respond in meaningful and adaptive ways.

Types of Computer Vision Tasks in HRI

There are various types of vision tasks essential for a robotic framework to perceive and interact with its surroundings in HRI. Each of these tasks focuses on a different aspect of vision and helps robots achieve specific capabilities in their interactions with humans.

1. **Object Detection and Recognition**: This task involves identifying and recognising objects in the robot's environment. Object detection usually involves developing an algorithm that, given an image or a sequence of images (e.g. from a video), can locate an object within the image. This is typically treated as a regression problem in machine learning, where the model predicts the coordinates of bounding boxes around the object. On the other hand, classification involves assigning the detected object to a specific category or label (e.g. "person", "chair") and is considered a classification problem in machine learning. Algorithms like *YOLO (You Only Look Once)* (Redmon et al. 2016) and *Faster R-CNN* (Ren et al. 2017) perform both tasks simultaneously—detecting objects and classifying them into predefined categories. In practice, object detection models learn to identify key features, such as edges, shapes, and textures, from visual data to locate and categorise objects, even under challenging conditions like poor lighting or partial occlusion. The successful implementation of object detection and recognition enables robots to interact with their surroundings more effectively. For example, a robot in a kitchen can detect a cup on a counter and correctly recognise it as a "cup", facilitating tasks like grasping or serving.
2. **Pose Estimation**: Pose estimation refers to the task of determining the positions and orientations of key points or joints on objects or human bodies in an image or video. In the context of HRI, pose estimation is primarily focused on detecting the

positions of human body parts such as the head, hands, and limbs, which are crucial for understanding gestures, movements, and intent. Unlike object detection, which treats objects as rigid entities, pose estimation requires the identification of flexible, articulated structures like the human body. Pose estimation is often addressed as a regression problem, where the model predicts the spatial coordinates of key points (e.g. joints in a human body). This is typically done by first detecting the person in the image and then using deep learning models such as convolutional neural networks (CNNs) to estimate the precise location of their key points. Popular models like *OpenPose* (Cao et al. 2021) or *DeepPose* (Toshev and Szegedy 2014) can estimate both 2D and 3D poses, providing robots with a detailed understanding of the orientation and movement of human subjects in real time. In HRI, pose estimation is important as it allows robots to interpret non-verbal communication cues, such as pointing, waving, or other gestures. For instance, a robot equipped with pose estimation can track a person's hand movements to identify when they are giving a command or manipulating an object. Moreover, 3D pose estimation provides robots with an understanding of the depth and spatial arrangement of human body parts, allowing them to navigate safely in shared environments or assist in tasks that require human-robot collaboration. Pose estimation has emerged as a crucial technique for converting more nuanced aspects of human interaction, as communicated through behavioural signals, into practical insights for HRI. Through precise monitoring of body components, such as the hand, body position, and object interactions, pose estimation technologies offer a holistic perspective on the manifestation of intentions in physical activities.

3. **Facial Recognition and Emotion Detection**: Facial recognition identifies or verifies a person based on facial features, while emotion detection interprets emotional states from facial expressions. In HRI, these tasks enable robots to personalise interactions and respond empathetically, improving engagement. Facial recognition is mostly a classification problem where models like *FaceNet* (Schroff et al. 2015) or *DeepFace* (Taigman et al. 2014) extract facial features and match them to known identities. Emotion detection, framed as a multi-class classification task, involves analysing expressions such as smiles or frowns to infer emotions like happiness or anger. Models such as *Affectiva* (Kulke et al. 2020) or *OpenFace* (Baltrušaitis et al. 2016) are commonly used to track facial features and detect emotions in real time. For use in HRI, facial recognition can help robots personalise experiences by identifying users, while emotion detection can allow them to adapt their responses based on a person's emotional state. For example, in healthcare, a robot that detects signs of discomfort in a patient can offer reassurance or notify a caregiver.
4. **Gaze Estimation**: Gaze estimation involves determining where a person is looking, which is often important for understanding attention and intention in HRI. It involves tracking the direction of a person's gaze using facial landmarks, such as the pupil, head orientation, and sometimes the iris, to predict a person's point of focus. Gaze estimation is often tackled using a combination of machine learning algorithms and traditional computer vision techniques that process facial images to estimate gaze direction (Pathirana et al. 2022; Adebayo et al. 2024). Gaze

estimation can be approached as either a 2D or 3D problem. In 2D, the model predicts the gaze vector on a plane (e.g. a screen), where the projection of gaze direction intersects at a point on the surface, giving coordinates (x, y), which represent pixel coordinates. In 3D, the problem becomes more complex, as the algorithm predicts a 3D gaze vector to estimate the precise location of attention in a real-world environment, rather than on a flat surface. Modern methods often employ deep learning models to infer gaze from high-dimensional facial data. In HRI, gaze estimation helps robots infer where a human's attention is directed, which is important for tasks that involve joint attention or understanding a user's focus during collaboration (Adebayo et al. 2024). For instance, in assistive care, a robot can use gaze estimation to detect when a patient is looking at a particular object and respond accordingly. In other cases, gaze direction can be integrated into an HRI framework as a form of 'direction' for relative affordances, where the robot uses the gaze information to assess potential actions or interactions based on what the user is focused on. For example, a robot in a collaborative setting might use gaze estimation to identify which tool a human is focusing on, adjusting its behaviour to assist in the task by handing over or manipulating the tool.

5. **Action Recognition**: There are several aspects of action recognition, and the problem is quite broad. However, in the context of HRI, action recognition often specifically refers to Human Activity Recognition (HAR). HAR focuses on identifying and classifying human activities, such as walking, sitting, or manipulating objects, and is central to understanding human behaviour in HRI—often the driving force of human-robot interaction, as identifying human actions enables robots to respond intelligently and proactively in collaborative settings. In HAR, various computer vision techniques are used to detect and analyse movements over time. For example, 3D Convolutional Neural Networks (3D CNNs), Recurrent Neural Networks (RNNs), and Long Short-Term Memory networks (LSTMs) are frequently employed to capture temporal dynamics, motion patterns, and long-term dependencies in video streams (Jobanputra et al. 2019; Shuchang 2022). By understanding human activities, robots can anticipate the next steps in a task and adjust their behaviour accordingly. All of the aforementioned CV tasks, such as object detection, pose estimation, facial recognition, emotion detection, and gaze estimation, can be combined to form a comprehensive solution for HAR. By integrating these CV tasks, robotic frameworks can gain a holistic understanding of human behaviour, improving the robot's ability to assist, interact, and collaborate with humans seamlessly (Jobanputra et al. 2019; Shuchang 2022). For example, in a healthcare setting, a robot could use gaze estimation to understand patient attention, pose estimation to track body movements, and emotion detection to assess the patient's emotional state, all contributing to a robust HAR framework.

1.2.2 Proprioception in Robots

Proprioception refers to a robot's ability to sense the position, movement, and orientation of its body or parts in space without relying on external visual feedback. In humans, proprioception enables us to perform actions such as walking, grasping objects, or maintaining balance without consciously thinking about the precise movements. For robots, proprioception is achieved through internal sensors, such as joint encoders, gyroscopes, and accelerometers, which help monitor the robot's posture, joint angles, and motion. This internal sensing ability is critical for enabling robots to maintain body awareness during tasks that involve physical manipulation or complex movements. For instance, in dyadic interactions where a robot may need to lift an object while simultaneously coordinating with a human, proprioception allows the robot to adjust its grip and maintain balance in real time.

Proprioception also complements external sensory inputs, such as vision or tactile feedback, providing the robot with a more holistic understanding of both its own body and the environment with which it interacts. This capability is essential for fine-tuning the robot's movements, avoiding collisions, and performing tasks with precision, particularly in dynamic environments or shared workspaces.

1.2.3 Natural Language Processing

Natural Language Processing (NLP) is a set of computational techniques that allow machines or systems to understand, interpret, and generate human language. In the context of HRI, NLP is often seen as an integral part of human interaction, especially in humanoid robots, where intuitive and natural communication is central to the interaction, contributing to smoother and more dynamic exchanges.

NLP in HRI encompasses several aspects, including speech recognition, language understanding, and language generation. The use of NLP enables robots in dyadic HRI to comprehend human speech, understand context, and respond meaningfully. This allows robots to go beyond simple, predefined commands, creating more fluid and natural interactions that closely resemble human-to-human communication.

Types of NLP Tasks in HRI

1. **Speech Recognition**: Speech recognition is the algorithmic process of converting spoken language into text. It forms the foundation for verbal communication between humans and robots. Advanced systems, like *DeepSpeech* (Hannun et al. 2014) or those based on deep learning models, are trained on large datasets to recognise various accents, speech patterns, and colloquial expressions. Accurate speech recognition is critical for robots to understand spoken commands and respond appropriately.
2. **Natural Language Understanding (NLU)**: After converting speech into text, the next challenge is for robots to grasp the meaning behind the words. This is where

NLU comes in. NLU enables robots to identify the intent, context, and meaning of what is being said. This involves recognising key information, detecting emotions, and understanding the purpose of the conversation. Advanced models like *BERT* (Devlin et al. 2019) or *GPT* (Radford et al. 2018) allow robots to follow complex conversations and respond appropriately in real time.
3. **Natural Language Generation (NLG)**: Once a robot understands what has been said, it needs to reply in a way that makes sense. NLG is the process of generating human-like responses. In conversational settings, NLG helps robots craft replies that are clear and relevant to the context. For instance, in healthcare, a robot might use NLG to provide clear instructions to a patient or offer reassuring words during an interaction.
4. **Dialogue Management**: Conversations between humans and robots often require multiple exchanges. Dialogue management helps robots keep track of the conversation, remember what has been said, and decide how to respond. By managing the flow of dialogue, robots can handle longer or more complex conversations, whether assisting a customer or providing ongoing care in healthcare. Techniques like machine learning allow robots to adapt to changes in the conversation and ensure that the interaction stays on track.

1.2.4 Artificial Intelligence (AI)

In dyadic HRI, the ability to perceive, understand, and respond to human actions and communication is made possible through a series of interconnected systems, all driven by algorithms. Whether it is computer vision helping a robot interpret visual cues (Qi et al. 2024) or NLP enabling robots to understand and generate human speech, algorithms are the backbone of these capabilities (Obaigbena et al. 2024; Soori et al. 2023). These algorithms work together, enabling robots to adapt to their environment, predict human intentions (Hoffman et al. 2024), and make decisions in real time (Trends in Control and Decision-Making for Human–Robot Collaboration Systems 2024). However, before these advanced machine learning techniques became central to robot intelligence, traditional rule-based systems were the foundation of early HRI (Leusmann et al. 2024; Obaigbena et al. 2024; Dahiya et al. 2023).

In traditional approaches, robots were programmed with fixed rules to follow specific instructions based on predefined inputs (Ahmadzadeh and Masehian 2015). While effective in controlled environments, these systems struggled with the complexity and unpredictability of real-world human interactions (Sarker et al. 2024). The need for more adaptable, intelligent robots revealed the limitations of these approaches, paving the way for the integration of AI (Kwasny and Faisal 1990; Obaigbena et al. 2024).

AI enables robots to process information, make decisions, and learn from their interactions with humans (Lemaignan et al. 2017; Doncieux et al. 2022). Rather than adhering to a fixed set of rules, AI allows robots to adapt to new situations and enhance their performance over time (Nolfi 2022). This adaptability is important in HRI, where human behaviour is unpredictable, requiring robots to be flexible and responsive. Key AI techniques, such as machine learning and deep learning for recognition and prediction, as well as reinforcement learning for experiential learning, empower robots to understand their environment and make intelligent choices in real time (Soori et al. 2023; Tang et al. 2024).

Core AI Technologies in HRI

1. **Machine Learning (ML)** is fundamental to the development of adaptive robotic systems in HRI. Rather than relying on pre-programmed responses to specific inputs, ML algorithms enable robots to detect patterns in data, learn from observations, and improve their interactions with humans over time (Semeraro et al. 2023). For instance, a robot can be trained to recognise various human gestures, emotions, or objects by processing large datasets, allowing for more intuitive and fluid interactions (Machine learning for human emotion recognition 2024; Hussain et al. 2014). In HRI, ML enhances a robot's capability to interpret and anticipate human actions, leading to more effective collaboration and contextual responsiveness (Obaigbena et al. 2024; Ahmad et al. 2017; Churamani et al. 2022).
2. **Deep Learning (DL)** is a subfield of machine learning that uses deep neural network models (i.e. neural networks with many layers) and associated training algorithms to learn from and interpret large amounts of unstructured data. DL models can perform much more complex processing than traditional machine learning models and excel in high-dimensional tasks such as visual recognition, natural language processing, and speech understanding. In HRI, DL allows robots to perform advanced perception tasks, such as recognising faces, understanding speech nuances, and interpreting human body language (Machine learning for human emotion recognition 2024; Hussain et al. 2014). These capabilities are critical for enabling robots to engage in more sophisticated, human-like interactions. DL's ability to model nonlinear relationships in data makes it essential for tasks that require high levels of abstraction and generalisation.
3. **Reinforcement Learning (RL)** allows robots to autonomously learn optimal behaviours through a process of trial and error. In RL, a robot is not explicitly told how to act; instead, it explores various actions within its environment and receives feedback based on the outcome of those actions. This approach is especially beneficial in dynamic, collaborative settings where robots must adapt to human behaviour in real time. By learning from experience and feedback, robots can refine their strategies, improving performance in tasks such as assisting a human operator or executing complex, adaptive movements in shared workspaces (Sharkawy and Koustoumpardis 2022).
4. **Generative AI (GenAI)** refers to a class of algorithms that can create new data based on patterns learned from existing datasets. In HRI, GenAI introduces novel capabilities that extend beyond the reactive nature of traditional AI systems.

Robots equipped with GenAI can synthesise human-like speech, generate realistic images, and simulate future actions, allowing for more dynamic and personalised interactions. By leveraging GenAI, robots can engage in richer dialogue, create tailored responses based on individual preferences, and simulate environments that help anticipate user needs (Obaigbena et al. 2024; Human-machine collaborative decision-making 2024).

1.2.5 Multimodal Integration

Multimodal integration refers to the fusion of sensory inputs from different modalities—such as vision, audio, haptics, and proprioception—enabling robots to form a holistic understanding of their environment and interactions. In HRI, it is essential for robots to process multiple simultaneous cues to ensure context-aware and adaptive responses.

Human communication inherently involves multimodal signals, where visual, auditory, and tactile information is combined to generate meaningful interactions. Robots that integrate and interpret inputs from diverse sensory channels are better equipped to replicate human-like behaviour, respond to non-verbal cues, and function effectively in dynamic environments.

Two primary approaches are commonly employed for multimodal integration in robotics. *Feature-level* fusion integrates data from different sensory sources at an early stage, allowing for joint learning across modalities (Sun et al. 2016; Roychoudhury et al. 2023). This method is often applied in tasks like facial emotion recognition, where both visual and auditory data are combined to enhance predictive accuracy. In contrast, *Decision-level* fusion processes each modality independently, and then merges the individual outcomes to make a final decision (Tsiourti et al. 2019; Roychoudhury et al. 2023; Sun et al. 2016). This approach provides flexibility by allowing robots to adapt even when one sensory input is compromised or unavailable.

Recent advancements in deep learning have significantly enhanced multimodal integration, particularly through the development of neural networks capable of processing and learning from diverse inputs such as visual, auditory, and proprioceptive data (Chung et al. 2019; Chen et al. 2021; Su et al. 2023; Gong et al. 2024). These systems facilitate more sophisticated interpretation of human actions, speech, and intentions, allowing robots to perform tasks such as collaborative assembly or assistive care with greater precision and responsiveness.

Multimodal integration is fundamental to HRI, as it enables robots to perceive and interpret their environment comprehensively. This capability is critical for developing adaptive and flexible robotic behaviours, enhancing their effectiveness in various domains, from industrial settings to healthcare (Roychoudhury et al. 2023).

1.2.6 Digital Twins

The term *Digital Twin* (DT) has become a key concept in emerging technology, with its origins rooted in computational science and engineering. A digital twin is a virtual representation that mirrors a physical entity or system, not as a static replica, but as a dynamic, real-time emulation that captures the complexities and interconnections of its real-world counterpart. Initially introduced in aerospace and manufacturing, the DT concept has evolved far beyond these domains, now encompassing a broad spectrum of applications—ranging from healthcare and urban planning to advanced robotics and artificial intelligence systems. This widespread adoption reflects the versatility of DTs, as they offer unparalleled capacity for detailed, dynamic simulations, allowing for in-depth analysis, optimisation, and predictive modelling. As a result, DTs have become a pivotal tool and enabler across numerous industries, transcending traditional boundaries.

The impact of digital twins is particularly evident in the field of HRI. Their introduction has been transformative, enabling the creation of sophisticated virtual models that encapsulate the complexities of both robots and humans, as well as their interactions in the physical world (Gallala et al. 2022; Lu et al. 2023). These models provide crucial insights into human behaviours, responses, actions, reflexes, and preferences when interacting with robots. Digital twins allow for the simulation and testing of diverse interaction scenarios, aiding in the development of robotic systems that are more responsive and attuned to human needs and emotions (Baratta et al. 2023; Lu et al. 2023; Miller and Spatz 2022; Ramasubramanian et al. 2022).

Furthermore, the integration of digital twins in HRI accelerates the creation of advanced, human-centric, and empathetic robotic systems. By simulating human-robot interactions in controlled virtual environments, researchers and engineers can address challenges more efficiently and explore new frontiers in robotics. Digital twins also improve robots' adaptability to human behaviours, enhance their decision-making capabilities, and refine their ability to interpret and respond to human emotional states. This integration contributes to the development of more advanced and empathetic robotic systems, better equipped for nuanced, human-centric interactions (Ramasubramanian et al. 2022; Malik and Brem 2021; Barresi et al. 2022; Asad et al. 2023).

1.3 Stages of Dyadic Interaction

Dyadic Human-Robot Interaction (D-HRI) is inherently reciprocal, characterised by a dynamic exchange of signals, interpretations, and responses between the human and the robot. As illustrated in Fig. 1.2, these interactions involve several distinct stages, each contributing to the robot's ability to perceive human behaviour, infer intentions, make decisions, and act accordingly. The first stage, *Perception and Detection*, involves the robot's ability to sense its environment and the human's actions; this

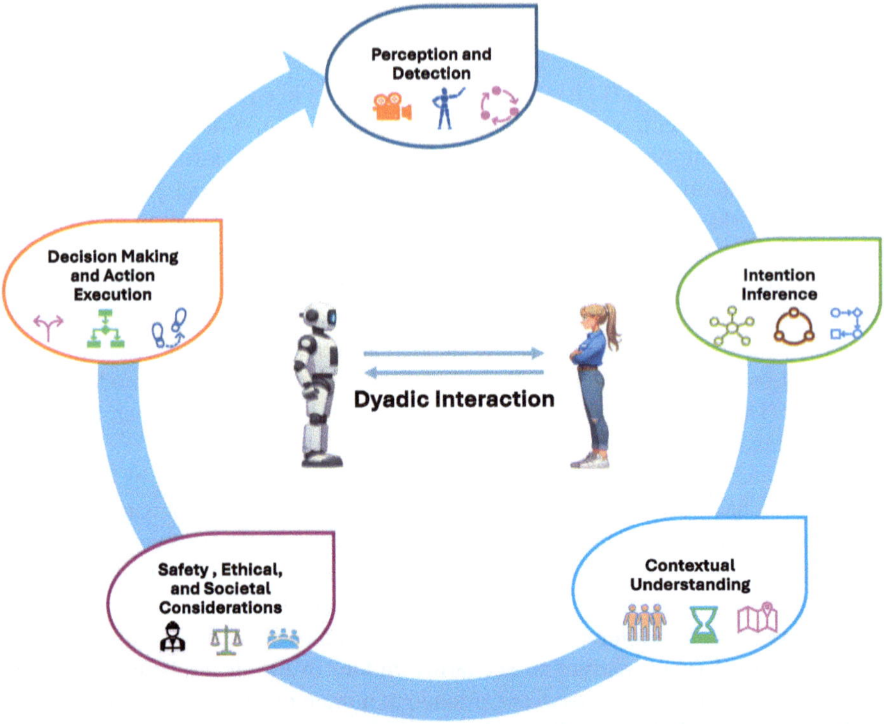

Fig. 1.2 Stages of dyadic human-robot interaction: perception, intention inference, contextual understanding, safety and ethical assessments, and decision-making and action execution. These are illustrating as forming a continuous feedback loop reflecting the continuous adaptation which is central to dyadic interaction

is then followed by *Intention Inference*, where the robot seeks to predict or understand the human's goals or next moves based on visual or behavioural cues; once the robot has inferred these intentions, *Contextual Understanding* allows it to situate this information within the broader social, spatial, or temporal context of the interaction. This is then followed by consideration of *Safety and Ethical aspects*, which applies across all stages, to ensure that the robot's behaves responsibly and maintains human trust and safety. Finally, in *Decision-Making and Action Execution*, the robot selects the most appropriate action based on its analysis, adjusting dynamically as the interaction unfolds. These stages, which form a continuous loop, are the foundation for enabling robots to engage in adaptive and meaningful collaborations with humans across various contexts.

1.3.1 Perception and Detection of Human Cues in Dyadic HRI

The first stage of dyadic interaction involves the robot's ability to detect and interpret human cues. This encompasses both visual and non-visual signals that enable robots to assess human behaviour and their surrounding environment. In human-centric frameworks, robots primarily rely on computer vision systems to detect gestures, facial expressions, body posture, gaze direction, and object interactions. For example, detecting hand movements during object manipulation can inform the robot of ongoing tasks (Adebayo et al. 2022). Additionally, non-visual cues, such as the tone and prosody in human speech, provide non-verbal auditory information that can signal emotional states or indicate urgency.

Just as human vision serves as a gateway to understanding the world, facilitating communication, and fostering social bonds, vision systems in robots play a similarly critical role. In dyadic HRI, these systems enable robots to perceive human gestures, emotions, and actions in real time, thus contributing to smoother, more contextually aware interactions.

Early vision-based approaches in HRI focused on object localization, identification, and tracking. These traditional methods, often utilising techniques like edge detection, corner detection, and contour identification, allowed robots to locate objects and infer trajectories, with low-level decision-making being guided by rule-based logic. While computationally efficient, these methods struggled in dynamic and complex environments due to their rigidity and limited adaptability (Canny 1986; Harris and Stephens 1988).

With the advent of machine learning, and, more recently deep learning, the field of vision-based systems in HRI has undergone significant transformation. Today, most state-of-the-art perception modules in robotic systems leverage machine learning techniques for more sophisticated analysis of visual cues, particularly in dyadic interactions where non-verbal communication plays a pivotal role.

Modern vision-based systems often integrate multimodal data, combining visual information with tactile or auditory inputs to enhance the robot's comprehension of its environment and the human actors it interacts with. For instance, advanced dyadic frameworks now enable robots to identify individuals, track their gaze to assess attention levels, and integrate this with speech recognition to generate appropriate responses during conversations. These capabilities allow robots to engage in more meaningful and adaptive interactions, pushing the boundaries of human-robot collaboration (Roychoudhury et al. 2023).

1.4 Intention Inference in Dyadic HRI

One of the key challenges in dyadic human-robot interaction (HRI) is enabling robots to infer human intentions, a problem commonly referred to as human intention inference (Adebayo et al. 2022). This process involves identifying the current task and predicting subsequent actions. While humans perform this instinctively due to years of social and cognitive development, for robots, it remains a highly complex task. One area of significant research interest is vision-based intention inference, which relies on interpreting visual cues such as gaze direction, hand movements, and body posture (Adebayo et al. 2022, 2024; Asad et al. 2023).

1.4.1 Vision-Based Intention Inference

In general HRI, including dyadic interaction, intention inference has evolved through various approaches, ranging from simple probabilistic models to sophisticated deep learning techniques. Early attempts at vision-based intention inference were challenged by the inherent ambiguity in human behaviour, making deterministic models inadequate for capturing human unpredictability. To address this, Inagaki et al. (1995) proposed a probabilistic approach, suggesting that human behaviour's inherent stochasticity demands models capable of handling uncertainty. This approach involved analysing behavioural indicators like gaze, body posture, and motion dynamics, and interpreting them using Fuzzy Logic within structured environments.

Since then, the integration of probabilistic models and machine learning has improved the accuracy of intention inference. For instance, the Intention-Driven Dynamics Model (IDDM), which simulates human movement patterns to deduce intentions, has shown promising results in addressing real-time constraints and requirements for adaptability in applications, such as robot table tennis and action recognition in interactive humanoid robots (Wang et al. 2013). Another method utilises Hidden Markov Models (HMMs) to predict human behaviour by learning intentional movement patterns online, enabling real-time inference (Vasquez et al. 2008). This has been extended using inverse reinforcement learning (IRL) to derive utility functions from observed behaviours, enhancing models for goal-oriented actions (Ziebart et al. 2008). Bayesian models derived from cognitive science have been utilised to deduce objectives and policies of human agents, thereby facilitating the prediction of human behaviour in situations similar to those previously observed or modelled (Baker et al. 2011).

The use of Gaussian Process (GP) models in intention inference, particularly in gaze-following behaviour, offers a valuable method for representing and learning about action dynamics driven by goals (Lawrence 2003). These models have significantly contributed to solving complex tasks in various domains such as autonomous driving, robot control, and human crowd navigation (Koppula and Saxena 2014). This illustrates the growing trend towards integrating intention inference with decision-making processes, enabling robots to better understand and anticipate human actions. However, as Wang et al. (2013) suggests, some approaches opt to separate these aspects, allowing for better handling of high-dimensional data streams and enhancing the focus on predicting human intentions. This separation ensures that the accuracy of intention prediction is not hindered by the challenges of robot decision-making, enabling more effective human-robot collaboration.

1.4.2 Progress in Gaze-Based Intention Inference Estimation

Building on foundational studies, recent advancements in vision-based intention inference have significantly enhanced the field of Human-Machine Interaction (HMI). In particular, gaze-based intention estimation has emerged as a pivotal area of research, yielding valuable insights into the psychological underpinnings of visuomotor control and its practical applications in teleoperated and assistive robotic systems (Belardinelli 2024). This research delineates a hierarchical categorization of intentions into distal, proximal, and motor, each functioning at distinct levels of action representation and time scales. Such an understanding has informed and influenced the development of contemporary gaze-based prediction techniques (Tatler et al. 2010; Pacherie 2008).

Furthermore, the intricate relationship between visual attention and motor intention has garnered increased scrutiny, with early findings suggesting that attention serves as a precursor to purposeful movement (Rizzolatti et al. 1987). This relationship is vital, as the visual system leverages prior experiences and expertise to anticipate the information necessary for effective motor planning (van de Laar et al. 2022; Itti and Baldi 2009). By understanding these dynamics, researchers are better equipped to design robotic systems that can accurately interpret and respond to human intentions in real time.

The study conducted by Onnasch et al. (2023) investigated the influence of abstract anthropomorphic gaze cues in facilitating attentional shifts and showed that a robot's eye movements can serve as effective sources of directional information. While this supports the cooperative eye hypothesis (Tomasello et al. 2007) and suggests that abstract anthropomorphic eyes may lead to quicker responses related to reflexive gaze cueing, it does not align with a human-centric approach to HRI. Instead, the emphasis is placed on responses that consider human intentions and attention as integral components of the robot's decision-making processes. Nevertheless, these discoveries play a crucial role in advancing HRI systems, particularly in industrial settings, by integrating human-centredness through human intention and fulfilling design criteria such as functionality and ease of implementation (Onnasch et al. 2023).

Moreover, the concept of action emerging from a fusion of plans and motor programs encapsulates the intricate essence of purpose. This perspective enriches the ongoing discussion surrounding gaze-based intention estimation by proposing that the eyes not only perceive but also signal future physical involvement. The integration of cognitive models with empirical data is pushing the boundaries of understanding how intention and attention converge to direct human-machine interactions.

The significance of visual and gaze systems in translating intention into motor action is well acknowledged, establishing gaze control as a primary indicator during the action preparation process. This assertion aligns with the premotor theory of attention (Rizzolatti et al. 1987), which posits that spatial attention results from enhanced spatial processing at potential action sites. This concept clarifies the dynamic interplay between attention and action, particularly highlighting the processes of selection-for-perception and selection-for-action (O'Sullivan et al. 2015). The visual system's role in optimising goal-directed actions is emphasised, aiming to maximise information acquisition and minimise uncertainty by leveraging prior experiences and knowledge to predict essential information required for such actions (van de Laar et al. 2022; Itti and Baldi 2009).

Recent advancements in head-mounted eye-tracking technology have enabled ecological investigations into eye-hand coordination, shedding light on the time-dependent relationship between gaze and object manipulation. These developments enhance our understanding of how gaze anticipates and guides manual movements (Johansson et al. 2001; Hayhoe et al. 2003; Pelz and Canosa 2001; Land 2009).

Consequently, the application of gaze-based intention estimation has become vital in enhancing our comprehension of how deliberate visual attention influences and directs physical interactions within human-machine interface contexts (Tatler et al. 2010).

1.4.3 Pose Estimation and Its Role in Understanding Intentions

The inherent dynamism of human activity presents unique challenges in estimating posture, particularly in environments with multiple individuals. Pose estimation methods generally follow either a top-down paradigm, where regions with individuals are identified before extracting features for pose estimation, or a bottom-up paradigm, which identifies target outputs first and then associates them with different individuals. The effectiveness of these paradigms relies on the careful design of posture encoder and decoder systems, which must adeptly navigate the complexities of distinguishing body parts, especially in crowded or obscured environments. Recent advancements have explored structural models, multi-scale feature fusion, multistage pipelines, refinement techniques, and multi-task learning to enhance accuracy and robustness in these challenging scenarios (Liu et al. 2022).

Traditionally, achieving highly accurate vision-based human position assessment was considered a formidable task. However, significant progress in deep learning has markedly advanced the field of human pose estimation, facilitating more accessible and precise capture and analysis of human movements. *Mediapipe*, a state-of-the-art pose estimation model developed by the Perception team at Google Research, exhibits impressive performance and operates in real time on relatively modest conventional computing platforms (MediaPipe 2024). Mediapipe's hand pose estimation is built on two cutting-edge models: Blazepalm, a variant of the Blazeface model tailored for palm detection in any input image, and the hand landmark model, which analyses the output of the palm detector to generate 21 hand landmarks. The training of Mediapipe hands utilised a large and diverse dataset, incorporating synthetic, real, and in-the-wild images. The model demonstrates robustness, exceptional speed, and proficiency in hand detection even in low-light conditions. Both lightweight and heavyweight iterations have been developed, with the former intended for edge and mobile devices, and the latter for high-performance servers.

Additionally, frameworks such as *OpenPose* (Cao et al. 2021), which employs Part Affinity Fields (PAFs) to accurately analyse the posture of multiple individuals in a single image, have significantly contributed to the advancement of pose estimation technology. This capability enables real-time multi-person 2D pose estimation inference. Enhancing these functionalities is *YOLOv8*, the latest version in the YOLO (You Only Look Once) (Redmon et al. 2016) series of object detection models, renowned for its remarkable speed and precision. Developed by Ultralytics, YOLOv8 offers an open-source model that integrates object detection and pose estimation functionalities, among other tasks. This versatility positions it as a fundamental model for various applications requiring knowledge of human posture and movement (Pose—Ultralytics YOLO Docs 2024). Ongoing advancements in model architecture and feature extraction, including structural models, multi-scale feature fusion, and multi-task learning, significantly improve the accuracy of pose estimation, particularly in densely populated scenarios and intricate settings (Liu et al. 2022).

1.4.4 Multi-visual Cues for Human Intention Inference

While traditional approaches to intention inference often relied on singular visual cues—such as gaze, body posture, or hand movements and gestures—recent advancements have led to a shift towards multi-visual cue fusion techniques. These techniques aim to combine multiple visual signals to provide a richer and more accurate understanding of human intentions. The idea of fusing different visual features to infer human intention stems from the concept that no single cue can fully capture the intricacies of human intention. In social and collaborative tasks, multiple cues are often used simultaneously—whether consciously or subconsciously—to convey intent. Dyadic robots must, therefore, adopt a similar approach to perception in order to match the flexibility and depth of human understanding.

Several recent approaches have explored fusing two or more signals to infer intention. For example, in Trombetta et al. (2020), the authors proposed a method to enhance human-robot collaboration by fusing information from both hand motion and eye gaze to predict human inference. They modelled human intention as a motion profile converging towards a goal location, enabling robots to anticipate future actions. By using skeletal tracking for hand motion and pupil tracking for gaze, they applied a Kalman filter to the eye gaze data to estimate the gaze point while transforming a shared reference frame with hand data. To further improve prediction accuracy, the authors employed an interacting multiple model (IMM) filter for hand motion tracking, fusing this data with the gaze filter's model probabilities. Experimental results from multiple reaching tasks demonstrated that the fusion algorithm could accurately predict human intentions before tasks were completed.

Building on this work, Adebayo et al. (2022) expanded the concept of multi-visual cue fusion by integrating hand movement, gaze direction, and object interaction for human intention inference. Their Hand-Eye-Object tracking framework emphasised the importance of combining these visual cues for more accurate and efficient prediction. The method employed a bidirectional LSTM neural network to classify human intentions based on visual inputs. This neural network structure allowed the system to learn from the temporal dependencies inherent in sequential human actions, a critical factor for capturing intention. Compared to systems using only two visual cues—such as hand movement and gaze—Adebayo et al. (2022) demonstrated through experimental studies that their three-cue model significantly outperformed simpler alternatives in terms of prediction accuracy and robustness.

Although facial expressions are typically explored within the realm of emotion recognition and affective computing, they are increasingly being integrated into intention inference systems in HRI. Combining facial expressions with gaze and body movements may add another dimension to the prediction of human intentions, especially in dynamic social and collaborative environments. For instance, a smile or frown combined with hand gestures or gaze direction can indicate whether an interaction is positive, neutral, or negative. In Li et al. (2016), the fusion of facial expression recognition (FER) with gaze and hand movements was explored for emotion-driven collaborative tasks. Using CNNs for facial expression recognition and combining these outputs with gaze and hand movement tracking, they showed that the integration improved the prediction of a user's emotional state and underlying intentions in real-time collaborative settings. Systems utilising machine learning and probabilistic models can effectively fuse these multimodal signals to predict human intentions in real time (Jain and Argall 2019; Chen et al. 2019). For example, the GazeMD system proposed by Shi et al. (2021) focuses on detecting visual intentions based on gaze patterns, further enhanced by incorporating facial and body movement data. This holistic approach enables robots to engage in more meaningful interactions by aligning their actions with the inferred intentions of human users.

While significant progress has been made in integrating these modalities, challenges remain in achieving robust and reliable intention inference in dynamic environments. Factors such as varying lighting conditions, occlusions, and individual differences in expression and gaze behaviour complicate the interpretation of cues (Rudovic et al. 2019). Future research should focus on developing adaptive algorithms capable of learning from diverse interaction contexts to improve the accuracy of intention inference over time (Lu et al. 2024; Zhang et al. 2022). Additionally, exploring the ethical implications of these technologies, particularly in sensitive applications such as caregiving, will be crucial for their acceptance and effectiveness (Wiese et al. 2019).

1.4.5 Non-visual Cues

Although visual cues provide a perceptual foundation for robotic frameworks, creating a truly human-centric dyadic interaction system involves much more than visual perception alone. Research highlights that perception must be complemented by non-visual modalities to achieve more natural and effective interactions, especially in assistive and social robotics (Urakami and Seaborn 2023; Fiore et al. 2013). In humans, non-visual cues can be broadly classified into two categories: verbal non-visual cues and non-verbal non-visual cues (A review of verbal and non-verbal human-robot interactive communication 2024).

Verbal Non-visual Cues: Verbal non-visual cues encompass aspects of human communication that are not reliant on visual elements, such as tone of voice, pace, and volume. These features can convey emotions or intentions that go beyond the actual words spoken. For instance, the work by Jahshan et al. (2013) provided evidence of a strong association between auditory processing and the identification of affective prosody in humans, suggesting that both early and late auditory event-related potentials play a role in differentiating emotions based on voice tone. Further evidence of this complexity is shown in Grichkovtsova et al. (2012), where the tone of voice (e.g. motivational) was found to forecast listeners' reactions.

In the context of HRI, particularly dyadic interaction, it is important to build robots that can adeptly employ both verbal and non-verbal cues. Indeed, it may also be crucial for robots to have emotional expressions and prosody that resemble those of humans in order to enhance interaction (Baylor 2011). Robots can be trained to adopt specific tones corresponding to various motivational communications, with the potential to enhance human engagement and emotional resonance (Paulmann et al. 2019). The expanding body of research on affective computing further supports the integration of emotional recognition capabilities into robots (Wu et al. 2016). By incorporating emotional intelligence into robotic systems, their capacity to react appropriately to human emotional states is augmented, fostering more authentic and effective interactions.

Non-Verbal Non-Visual Cues: Non-verbal non-visual cues refer to other forms of communication that do not rely on speech or visuals, such as body language, gestures, and even spatial dynamics. In robotic systems, the implementation of such non-visual cues presents unique challenges and opportunities, as robots lack the innate ability to produce or interpret auditory and olfactory signals in the same manner as humans. However, advancements in technology have enabled robots to simulate and respond to these cues. Research has shown that body language and gestures significantly influence communication, even in robotic environments. For instance, Collins et al. (2015) highlighted that humanoid robots can proficiently express affective states through body language, illustrating that gestures emulated from human behaviour can convey emotions.

Earlier research by Obaid et al. (2012) supported this idea by investigating algorithms for real-time recognition of full-body motions, which augmented user interaction with robots. This demonstrates that robots can be designed to comprehend and mimic human-like movements, further enabling more natural interactions. The capacity of robots to comprehend and respond to human body language is essential for effective human-robot interaction. McColl et al. (2011) studied the importance of robots recognising human non-verbal communication, particularly upper-body language, to assess an individual's accessibility and receptiveness during encounters. This ability allows robots to adapt their behaviour according to the emotional states of their human counterparts.

The development of systems capable of real-time recognition and classification of body language is crucial for enhancing the social intelligence of robots (McColl and Nejat 2012). Research into robots' ability to convey emotions through body language remains a prominent field of study. Beck et al. (2010) introduced the notion of an *Affect Space*, enabling robots to combine various emotional expressions via bodily movements. This method underscores that robots can effectively portray emotions through body language alone, without needing facial expressions, conveying feelings such as joy or sorrow. Investigations in Zheng and Meng (2012) further emphasised the creation of gestures with semantic significance for humanoid robots, noting that body postures can serve as a substitute for facial expressions in transmitting emotional content.

Despite these advances, challenges remain in the efficient integration of non-visual inputs within robotic systems. Robots must be equipped with advanced sensors and algorithms capable of accurately reading human motions and body language. Luo et al. (2020) highlighted the need for automated recognition systems that can identify emotional physical expressions, which are crucial for robots to form substantive relationships. The incorporation of multimodal signals, such as combining body language and auditory cues, can further enhance the overall conversational efficacy of robots (Tsiourti et al. 2019).

1.5 Contextual Understanding

Beyond the interpretation of human intentions, robots must integrate information about their environment and the broader context in which interactions take place. This often involves understanding both spatial and social contexts, as well as how human actions relate to the overall interaction dynamics. In this context, the term "environment" extends beyond physical objects to include social dynamics, task goals, and ambient factors (Jahanmahin et al. 2022; Stock-Homburg 2022). While detection and intention inference often involve recognising and predicting subtle individual human actions, contextual understanding enables robots to adapt to the broader, evolving conditions of the interaction space, prioritising fluidity and facilitating cooperative exchanges (Scofano et al. 2024; Huettenrauch et al. 2006; Jahanmahin et al. 2022; Stock-Homburg 2022).

There are several key areas of contextual understanding as applied to HRI. Given that the focus of this study is on dyadic interaction, we will specifically examine spatial awareness and environmental context, social context awareness, and temporal context for contextual understanding. Additionally, we will explore applications of these concepts in real-world systems.

Developing proactive robots is crucial for enhancing the quality of HRI. Rather than merely reacting to human actions, proactive robots are designed to anticipate future actions and adapt their behaviours accordingly. This anticipation leads to more intuitive and seamless collaboration. For instance, a robot equipped with spatial awareness can anticipate human movements in a shared workspace and adjust its own actions to avoid conflicts and improve the quality of interaction. Similarly, social context understanding allows a robot to interpret non-verbal cues and adapt its role within a group or task, thereby improving teamwork fluidity. Temporal awareness further enables robots to synchronise their actions with human partners, ensuring timely task completion.

Together, these layers of contextual understanding—spatial, social, and temporal—are foundational to creating robots that operate proactively rather than reactively, while also enabling adaptive switching between reactive and proactive behaviours depending on the context. This ability to fluidly transition between modes of operation will foster more natural, human-like interactions.

The following sections will explore these critical components of contextual understanding in dyadic interaction, focusing on spatial awareness and environmental context, social context awareness, temporal context, task awareness, and their integration into real-world systems.

1.5.1 Spatial Awareness and Environmental Context in Dyadic HRI

In D-HRI, spatial awareness is crucial for enabling robots to navigate and interact effectively within shared environments. It pertains to a robot's ability to perceive and adapt to its physical surroundings, including the understanding of how the environment itself influences the interaction. This awareness ensures that robots can respond dynamically to changes in spatial configurations and anticipate how their movements will affect human collaborators. Spatial awareness can be viewed as the application of detection and intention inference to environmental understanding, as these perceptual and predictive capabilities inform the robot's spatial behaviours.

While detection focuses on identifying relevant features and objects within the robot's environment, and intention inference predicts the likely evolution of human actions within that space, spatial awareness builds on these processes. It integrates environmental data into a cohesive model that allows the robot to react to immediate stimuli, while also planning and adapting in anticipation of changes in the shared environment during collaboration. This continuous integration of perception and prediction helps ensure smooth, cooperative exchanges between humans and robots, making spatial awareness a cornerstone of effective dyadic interaction.

Managing Shared Space in Dyadic HRI: In dyadic HRI, effective management of shared spaces is crucial for ensuring task efficiency and smooth collaboration. However, this task, which comes naturally to humans, presents significant challenges for robots—especially dyadic robots. Unlike autonomous systems that typically operate independently, dyadic robots must coexist with humans in the same workspace, performing tasks that require high levels of coordination and adaptation to human movements. Collaborative robots (cobots), designed to work alongside human counterparts, aim to complement human actions rather than replace them.

Several research subthemes define the key requirements for the effective navigation of robots around humans in a shared workspace. These subthemes include maintaining appropriate interpersonal distance through proxemics, understanding the significance of objects and tasks via environmental semantic understanding, and anticipating human movements through spatio-temporal prediction. Spatio-temporal prediction, closely linked to human intention inference (as discussed in Sect. 1.4), supports smooth navigation in shared spaces by enabling cobots to align their movements with those of human partners.

Proxemics studies how people perceive and manage personal space, which in dyadic HRI translates into maintaining an appropriate interpersonal distance between the robot and the human. This ensures safety, comfort, and seamless collaboration. The proxemics theory, developed by Edward T. Hall, identifies four spatial zones—intimate, personal, social, and public—that influence social interactions between humans (CSISS Classics—Edward T. Hall 1966; Mead and Matarić 2017; Rios-Martinez et al. 2015). These zones have been adopted in HRI to develop robots that can navigate around humans while respecting personal boundaries (Mead and Matarić 2016). While the primary objective of integrating proxemics into robotic frameworks is often practical—such as ensuring safe and efficient navigation—there is a growing focus on socially aware robots. These robots adjust their positioning based on proxemic cues, allowing for more natural and non-intrusive interactions (Mead and Matarić 2017, 2016; Mead et al. 2011). For example, robots in healthcare or service settings that respect personal space tend to inspire more trust and are perceived as more reliable by human collaborators (Mead and Matarić 2016; Mead et al. 2011).

Beyond proxemics, the ability to perceive the environment from the human's viewpoint (referred to as *spatial perspective-taking*), is essential for predicting human movements and adapting robotic actions in real time. While similar to human intention inference, spatial perspective-taking focuses specifically on how a robot positions and adjusts itself relative to the human's viewpoint and ongoing tasks to ensure smooth coordination. Studies show that humans frequently employ spatial perspective-taking when interacting with robots, which in turn affects the quality of the interaction (Abrini et al. 2023; Xiao et al. 2021). The design and movement of robots also significantly influence human attention and engagement. For instance, robotic motion can capture social attention, affecting how humans perceive and respond to robots in various contexts (Li et al. 2015). In environments where robots must navigate among people, adhering to social norms regarding personal space is critical for fostering positive interactions (Bera et al. 2017; Truong et al. 2017).

In addition to spatial awareness through proxemics and perspective-taking, robots in dyadic HRI must develop environmental semantic understanding—the ability to interpret the function and significance of objects and areas within a workspace. This capability enables robots to navigate shared environments effectively, distinguishing between task-relevant and non-relevant objects or areas.

Semantic understanding extends beyond basic object recognition by incorporating contextual knowledge about how objects are used in specific tasks. For example, a robot assisting in a kitchen needs to understand that a stove is for cooking and should avoid placing non-cooking items there. This level of environmental understanding allows the robot to make more informed decisions, minimising disruptions to human workflows.

Rusu et al. (2008) introduced semantic mapping techniques that enable robots to associate objects with their functions. Integrating this semantic knowledge into a robot's navigation and interaction framework allows it to move through a workspace in a task-relevant way, minimising interference with human activities. Semantic mapping enhances a robot's ability to prioritise tasks based on environmental context, improving accuracy and reducing unnecessary actions.

Recent advances in deep learning and semantic segmentation further bolster environmental semantic understanding. Works such as Guastella and Muscato (2021), Ginerica et al. (2024) and Visca et al. (2021) demonstrated how robots could use deep learning to interpret complex, unstructured environments such as homes or hospitals. In these settings, the robot can identify task-relevant areas (e.g. medical stations) and avoid irrelevant ones unless required, thereby assisting more effectively without constant human input. This level of understanding enhances the robot's autonomy in dynamic environments, where human tasks and priorities may change over time.

Semantic understanding is especially valuable in scenarios that demand precision and context-sensitive behaviour, such as industrial or manufacturing environments. A robot equipped with this capability can prioritise objects or areas based on task urgency, allowing it to support human collaborators more effectively under shifting conditions.

By developing a robust understanding of the functional relationships between objects and areas, robots in dyadic HRI can navigate shared spaces more intelligently, offering meaningful support to human collaborators while minimising errors and inefficiencies.

Spatial Awareness and Environmental Adaptation with Digital Twins: A significant advancement in spatial awareness for robots is the integration of digital twin technology, which allows robots to interact with real-time, dynamic models of their environments. By continuously updating virtual replicas of the physical world through data streams from sensors like LiDAR, RGB-D cameras, and computer vision systems, DT enables robots to simulate and predict environmental changes, enhancing their real-time decision-making and behavioural adaptation.

For example, Gallala et al. (2022) proposed a DT framework that utilised Microsoft Hololens 2 and a KUKA IIWA robot to maintain real-time simulations of the work environment. In this framework, non-expert operators could interact with robots through augmented reality (AR) interfaces, dynamically reconfiguring robot actions and enabling efficient collaboration with minimal programming knowledge (Gallala et al. 2022; Liao and Cai 2024). While these works primarily focus on human-robot interaction through AR, the implications for dyadic HRI are significant. By continuously updating spatial awareness and anticipating human movements, DT technology enhances the ability of robots to navigate shared environments and adapt to changes in real time.

Thus, integrating DT into spatial awareness systems improves the robot's capacity to manage tasks, navigate shared spaces, and respond to dynamic human behaviours, facilitating smoother and more efficient dyadic interaction.

1.5.2 Social Context Awareness

Early research in HRI predominantly centred around task-specific robot behaviours, where social context awareness played a secondary role. In these early models, robots were primarily designed to respond to basic social cues using rule-based systems, often relying on a limited set of inputs such as speech recognition or gesture detection (Lutfi 2013). A major limitation of these approaches, however, was their inability to fully understand and interpret the subtleties of human social behaviour. This restriction significantly limited the robots' ability to engage in meaningful and fluid interactions.

Over the past two decades, there has been significant evolution in social context awareness within HRI. This growth can largely be attributed to advancements in artificial intelligence, particularly in machine learning and deep learning, as well as the incorporation of affective computing into human-machine system design (McColl et al. 2016). As a result, robots are now better equipped to understand and predict human behaviour in real-time social environments. Recent research has expanded beyond simple responses to individual cues, integrating body language, facial expressions, and the dynamics of group interactions to prepare robots for more seamless and intuitive engagements in social settings (McColl et al. 2016; Zaraki et al. 2017; Hong et al. 2017; Bremers et al. 2024).

Social Awareness and Human Roles: In human-to-human interactions, individuals frequently adopt roles such as leader, supporter, or peer, adjusting their behaviours accordingly. For socially aware robots, the ability to recognise and adapt to these roles is crucial in dyadic HRI. Social role recognition enables robots to identify whether a human is leading a task, assisting in its completion, or participating as an equal peer, allowing them to respond appropriately and maintain fluidity in collaboration.

Understanding and adapting to these roles helps robots provide the right level of support or follow-up actions. For instance, robots capable of adjusting based on leadership roles improve teamwork by automating task coordination and reducing cognitive load on human collaborators (Hoffman and Breazeal 2004). Role recognition frameworks integrated into robot decision-making have shown improvements in task efficiency and rapport between humans and robots, particularly in collaborative settings such as construction or assembly tasks (Churamani et al. 2022; Frederiksen 2021).

In healthcare, social role recognition is especially important, as robots need to adjust their behaviour depending on whether they are interacting with patients, caregivers, or medical staff to provide effective assistance (Powers et al. 2007; A review of verbal and non-verbal human-robot interactive communication 2024). This adaptability ensures smoother interactions and enhances the robot's ability to meet the unique demands of various roles in dynamic social environments.

Early methodologies for role assignment in dyadic HRI laid the groundwork for understanding how robots can efficiently collaborate with humans. These approaches often centred around predetermined roles, dictated by the task specifications, where the robot typically assumed a subordinate or assisting role to the human counterpart. In Jarrassé et al. (2014) a framework is introduced that adjusts a robot's level of assistance dynamically based on human engagement and contribution, underscoring the importance of reciprocal interactions in tasks like handshaking, which inherently involve mutual action. This framework highlights the necessity for robots to not only perform tasks but also to exhibit socially aware behaviours that reflect the nuances of human social interactions.

In Mörtl et al. (2012) static role assignments for physical collaboration in dyadic HRI is examined, with a focus on how human exertion and task performance metrics can be used to determine optimal assistance levels. While roles in such models may remain fixed, they emphasise the potential for real-time role adaptation based on input from the human collaborator. This flexibility allows robots to respond to varying human inputs and adjust their actions accordingly, promoting a more interactive and dynamic workspace.

Further exploring the concept of role assignment, Jarrassé et al. (2014) investigated collaborative motor actions, proposing that roles can either be pre-assigned based on the nature of the task or develop dynamically over the course of the interaction. This adaptability is particularly vital in scenarios requiring extended physical collaboration, where human and robot roles may shift as the task progresses.

While these studies primarily focus on dyadic interactions, Lima et al. (2022) extended the investigation to non-dyadic interactions, introducing the complexity of multi-agent environments that involve both human and robotic participants. This expanded perspective challenges traditional static role assignments, requiring more sophisticated approaches to social awareness in HRI. Lima et al.'s research suggests that as robots become integrated into more complex social environments, their ability to adapt to multiple roles and understand group dynamics will be critical for effective interaction.

The evolution of social role recognition in HRI has been significantly bolstered by advancements in machine learning, affective computing, and multimodal communication. These technologies are instrumental in enabling robots to better understand human behaviours and emotions, facilitating more natural and effective interactions.

One notable approach involves the use of facial expression recognition systems, which enable robots to discern human emotional states through visual cues. However, as highlighted by Chaminade and Okka (2013), challenges arise when robots attempt to recognise human faces due to the limited resemblance between robotic systems and human features. These challenges, including a robot's lack of human-like perceptual abilities, can impair the speed and accuracy of processing of social cues, necessitating additional cognitive resources for effective interaction.

Improvements in deep learning methodologies have enhanced the accuracy of facial expression recognition, as demonstrated by recent research (Jirak et al. 2022; Spezialetti et al. 2020). These advancements allow robots to better understand and respond to human emotions. Moreover, multimodal approaches that combine multiple sensory inputs—such as auditory cues, non-verbal gestures, and visual attention—are increasingly being employed. Admoni and Scassellati (2017) emphasise the role of non-verbal cues, particularly eye gaze, in expressing a robot's engagement and cognitive states during interactions. This becomes especially relevant in complex social situations, where robots must interpret both individual emotional expressions and group dynamics to adapt their behaviours accordingly (Liu et al. 2021; Chita-Tegmark and Scheutz 2021).

Machine learning algorithms play a critical role in refining these recognition techniques. Zhao (2023) highlights the importance of training algorithms to identify emotional states and adjust robotic behaviours to foster collaboration in social contexts. Similarly, emotion recognition frameworks, as discussed by Mohammed and Hassan (2021), allow robots to adapt their actions based on the emotional context of interactions, thus enhancing user trust and experience.

An essential aspect of social role recognition pertains to understanding gendered perceptions of robots. Neuteboom and de Graaf (2021) illustrate how a robot's appearance and its assigned roles can influence human perceptions and trust. This indicates that future HRI designs must consider these factors to improve societal acceptance of robots. Huber et al. (2014) further support adaptive role assignments that transcend traditional stereotypes, advocating for robots that can flexibly modify their roles based on user expectations and the social environment.

Use of Social Cues in Interaction: Social cues are signals—both conscious and unconscious—that humans used to convey information, emotions, intentions, and expectations during social interactions. These cues, which include verbal and non-verbal behaviours, play a crucial role in facilitating smooth communication between individuals. In the context of HRI, social cues encompass the behaviours, expressions, and gestures that robots must learn to interpret and respond to in order to engage effectively with humans.

Beyond understanding roles and interactions in dyadic settings, non-verbal communication is an essential element that a dyadic robot must not only interpret but potentially mimic. Robots capable of recognising and responding to gestures, facial expressions, and body language are better equipped to engage with humans naturally and effectively. This concept is grounded in human-to-human interaction studies, which consistently show that much of human communication is non-verbal, relying on subtleties in body language, facial expressions, and gestures to convey emotions, intentions, and social cues.

Facial expressions, in particular, are a key non-verbal cue widely adopted in HRI to enhance engagement, offering valuable insights into a person's emotional state. This is especially important in healthcare settings, where recognising emotions like discomfort or stress can significantly improve patient care. Similarly, other non-verbal cues, such as gestures and overall body language, can help robots better interpret human actions and intentions.

Additionally, gaze and eye movements provide another layer of non-verbal communication, offering indications of attention and intention. For example, gaze-tracking enables robots to manage conversational turn-taking and infer a user's focus during task-oriented interactions (e.g. Trombetta et al. 2020; Adebayo et al. 2022, 2024). Moreover, integrating proxemics (CSISS Classics—Edward T. Hall 1966) can help robots maintain appropriate physical space during interactions. This capability is crucial for ensuring comfort and trust, especially in sensitive environments like healthcare, where maintaining a respectful distance contributes to positive human-robot collaboration (Lindsay et al. 2024).

Group Dynamics and Collaborative HRI: Another important dimension of social context awareness in Human-Robot Interaction (HRI) is the understanding of group dynamics, particularly in collaborative settings. While this paper primarily focuses on dyadic interaction, an understanding of group dynamics becomes crucial when multiple humans and robots work together. In this context, group dynamics refers to the evolving roles, relationships, and communication structures between human and robot team members during collaborative tasks. Recent research has shown that effective collaboration between humans and robots in group settings requires robots to adapt to shifting roles and maintain clear communication to support team objectives (Clabaugh et al. 2019; Traeger et al. 2020; Hoffman and Zhao 2020). Integrating robots into human teams can significantly influence group cohesion and overall performance, underscoring the importance of designing robots that can seamlessly fit into existing social structures (Hoffman and Zhao 2020; David et al. 2022; Henschel et al. 2021).

Robots, expected to function as part of human teams, must be able to identify and adjust to the different roles humans naturally adopt during teamwork, such as task leader, supporter, or coordinator. Recognising these roles enables robots to dynamically adjust their behaviour, offering support or taking on responsibilities as needed. For example, when a human assumes the role of task leader, a robot might defer to them for decision-making, assist with supplementary tasks, or take proactive measures to ensure smooth task execution. Likewise, a robot can step in to offer assistance when it detects that a human is struggling or requires help.

Several approaches in HRI integrate group dynamics, involving either individual robots or multi-party systems designed to interact with multiple humans simultaneously (Żarkowski 2019; Kwak 2014; Söderlund and Natorina 2024). These approaches focus on navigating the complexities of social interaction by interpreting and adapting to social cues, such as group hierarchies and role-taking behaviours, including identifying leaders or supporters within a group (Ghazali et al. 2019; Kwak 2014). Multi-party dialogue systems, for instance, enable robots to maintain conversations with multiple participants, using natural language processing (NLP) and gesture recognition to manage turn-taking and social roles (Dahiya et al. 2023; Nguyen et al. 2023). Similarly, cooperative robotic frameworks that employ machine learning allow robots to dynamically adjust their actions based on their recognition of social roles and context (Clabaugh et al. 2019).

When designing robots for multi-party interactions, user expectations and the social context of the interaction must be considered. Research by Kuo et al. (2011) highlights that fulfilling user expectations through appropriate robotic expressions of internal states leads to more satisfying interactions. This aligns with findings from Louie et al. (2014), which argue that robots need to autonomously manage the collective needs of individuals within a group to facilitate successful interactions. Planning and scheduling activities to accommodate multiple users is a significant challenge that requires advanced algorithms and socially aware functionality in robots (Louie et al. 2014).

Communication strategies also play a crucial role in enhancing group dynamics in HRI. Non-verbal cues, such as gestures, body language, and facial expressions, can greatly influence how robots are perceived in social contexts, contributing to a sense of "aliveness" or social agency (Urakami and Seaborn 2023). This is particularly important in multi-party interactions, where a robot's ability to interpret and respond to non-verbal signals can enhance collaboration and understanding among users. Additionally, reciprocity in interactions, as explored by Sandoval et al. (2021), suggests that robots exhibiting reciprocal behaviours are perceived more favourably, fostering a more cooperative atmosphere in group settings.

The successful integration of these approaches requires a multidisciplinary understanding of emotions and personality dynamics in public demonstrations of robots (Gunes et al. 2019). Such insights can guide the design of robots that are not only functional but also socially acceptable and engaging in group contexts. Additionally, the development of frameworks for analysing human roles in HRI scenarios will aid researchers and designers in creating more effective multi-party interaction systems (Puig-Pey et al. 2022).

1.5.3 Temporal Context Awareness

When humans interact, they rely on a broader context of the tasks and goals, often guided by the timing and sequencing of events. This awareness allows them to understand not just the "what" of actions but also the "when"—the order and timing of occurrences—enabling them to synchronise their behaviours with their partner. This ability to perceive, interpret, and act based on the timing and sequencing of events during an interaction is known as *temporal context awareness*. Similarly, robots in dyadic Human-Robot Interaction (HRI) must develop temporal awareness to foster natural collaboration with humans. Just as humans adjust their behaviours in response to temporal cues, robots must align their operations with human partners to facilitate smooth, efficient, and goal-oriented collaboration. Temporal awareness, therefore, is a fundamental aspect of human-robot collaboration, especially when actions are interdependent and time-sensitive.

Addressing the challenge of encoding temporal awareness into a robot's framework can take various forms. One approach involves using *temporal prediction models*, which allow robots to anticipate the future state of the environment or their human partner's actions based on past and present observations. For instance, robots can be equipped with Long Short-Term Memory (LSTM) networks or Recurrent Neural Networks (RNNs) to track temporal sequences and predict human movements or adjust their own speed to intervene when necessary. These foundational neural network models excel at capturing dependencies between events over time and are especially important for maintaining the flow of interaction and ensuring tasks are completed efficiently.

In addition to capturing temporal dependencies through prediction models, robots can enhance temporal awareness by recognising contextual cues. *Contextual cue recognition* is crucial in situations where robots must detect and interpret temporal signals, such as pauses, delays, or changes in the pace of interaction. For instance, in conversational HRI, pauses in speech may signal that a human partner is awaiting a response, while the identification of corrective gestures may indicate that the robot needs to adjust its behaviour. Failure to recognise these cues could disrupt or delay tasks, making the interaction feel unnatural.

In environments like industrial automation or healthcare, where timing is critical, temporal context awareness can be enhanced through *task scheduling algorithms*. These algorithms help robots prioritise tasks based on their time sensitivity and the state of human collaborators. For example, in an assembly task, a collaborative robot (cobot) working alongside a human can use temporal context to determine when to assist, based on the human's progress in completing previous tasks, while avoiding workflow bottlenecks.

Furthermore, robots can encode temporal awareness through *learning from temporal feedback*. By leveraging frameworks like reinforcement learning, robots can continuously improve their timing and responsiveness based on real-time feedback from their environment and human partners. This approach allows robots to refine their timing strategies through repeated interactions, improving their ability to predict when to act, how quickly to move, and when to wait.

1.6 Decision-Making and Action Selection

Throughout this paper, we have explored various components of dyadic interaction—perception, intention inference, contextual understanding, and multimodal integration—that collectively enable a robot to gather, process, and interpret information in real time. These capabilities form the foundation of decision-making, guiding robots as they select appropriate actions in dynamic and often unpredictable environments. Although decision-making is inherently embedded in all stages of dyadic interaction, understanding it as a distinct function is vital for building systems that can seamlessly adapt, learn, and operate in complex real-world settings.

Early Approaches to Decision-Making

Initial decision-making models in robotics relied heavily on rule-based systems, where robots executed predefined commands and followed simple logic trees. While effective in static environments, these models lacked the adaptability necessary for dynamic HRI. One of the first formal approaches was the Finite State Machine (FSM), where robots transitioned between different states based on specific triggers (Iovino et al. 2023). FSMs operate by restricting the robot to one state at a time, transitioning only when prompted by predefined inputs (State Machines vs Behavior Trees 2024). Though FSMs are relatively straightforward to implement, they struggle with scalability in complex, collaborative scenarios where robots must respond to unpredictable environmental changes (State Machines vs Behavior Trees 2024). This limitation rendered FSMs inadequate for real-time, scalable interactions, especially in human collaboration contexts were unpredictable behaviours frequently arise (Iovino et al. 2023; Wang et al. 2021).

As robotics advanced, alternative decision-making frameworks such as Behaviour Trees (BTs) emerged. Unlike FSMs, BTs offer modularity and flexibility, allowing robots to switch dynamically between tasks and manage more complex, nonlinear decision-making processes (State Machines vs Behavior Trees 2024; Iovino et al. 2024). These improvements enabled robots to better handle the demands of dynamic and collaborative environments, marking a significant step forward in HRI.

Probabilistic Decision-Making Models

While BTs offered greater flexibility than FSMs, both models still operated within largely deterministic frameworks. As robots began engaging in more complex and dynamic environments, particularly in collaboration with humans, the limitations of such deterministic approaches became increasingly apparent. Deterministic models struggled to handle uncertainties arising from unpredictable human behaviour, incomplete knowledge of the environment, or noisy sensor data. To address these challenges, probabilistic decision-making models, such as Markov Decision Processes (MDPs) and their extension, Partially Observable Markov Decision Processes (POMDPs), gained prominence.

Unlike FSMs and BTs, MDPs enable robots to evaluate various possible actions based on predicted outcomes and their associated probabilities (Lauri et al. 2023; Understanding the Markov Decision Process (MDP) n.d.). This probabilistic approach allows for more adaptive and flexible decision-making in uncertain environments. However, while MDPs provided greater adaptability than FSMs, they introduced significant computational challenges, particularly in complex HRI tasks that involve large state spaces and continuous interaction (Lauri et al. 2023; Understanding the Markov Decision Process (MDP) n.d.).

A more sophisticated variant, the Partially Observable Markov Decision Process (POMDP), addressed the challenges of incomplete or noisy information—a common issue in human interaction. Unlike MDPs, which assume full knowledge of the environment, POMDPs enable decision-making with partial observations, making them particularly valuable in tasks involving uncertain or ambiguous human behaviours (Lauri et al. 2023; Zacharaki et al. 2021). For example, in scenarios where robots have limited knowledge of human emotions, intentions, or physical states, POMDPs allow for more robust decision-making.

While POMDPs offer flexibility and resilience, their computational complexity increases significantly with larger state spaces and longer planning horizons, leading to challenges commonly referred to as the "curse of dimensionality" and the "curse of history" (Lauri et al. 2023). As a result, despite their advanced capabilities, the computational demands of POMDPs limit their widespread use in some real-time systems.

Time-Sensitive Decision-Making

In recent developments, Drift Diffusion Models (DDMs) have been applied to decision-making in time-sensitive scenarios. DDMs conceptualise decision-making as a process of accumulating sensory evidence until a decision threshold is reached, making them particularly useful in fast-paced environments like collaborative assembly, where both rapid response and precision are crucial (Fudenberg et al. 2020). In robotics, the use of DDMs enables systems to optimise decision times based on real-time input while maintaining accuracy, enhancing efficiency in high-pressure environments. For example, in collaborative assembly tasks, robots utilising DDMs can quickly adapt to changing conditions, making precise yet timely decisions that improve task performance and human satisfaction (Myers et al. 2022).

However, implementing DDMs requires careful calibration to ensure decisions are made within optimal time frames, avoiding delays or errors. This approach has proven effective in scenarios that demand a balance between speed and accuracy, such as patient care or precise mechanical assembly (Sajko and Babič 2024; Malaiya

2024). By modelling the trade-off between these factors, DDMs allow robots to dynamically adjust their decision-making processes, optimising for the specific requirements of the task at hand.

Machine Learning and Reinforcement Learning

While DDMs excel in time-sensitive decision-making tasks by balancing speed and accuracy, the broader scope of machine learning—particularly through Reinforcement Learning (RL) and its extensions like Deep Reinforcement Learning (DRL)—has empowered robots to make swift decisions while continuously learning and adapting in complex, dynamic environments. Recent shifts in decision-making models have highlighted the widespread adoption of machine learning, especially RL, which enables robots to learn optimal strategies through trial and error, receiving feedback in the form of rewards or penalties. This approach has proven effective in allowing robots to adapt dynamically to new tasks and environments (Zhang et al. 2021; Gronauer and Diepold 2022).

DRL, an extension that incorporates deep neural networks, has been pivotal in addressing complex decision-making problems in Human-Robot Interaction (HRI), such as the simultaneous interpretation of speech and gestures. DRL enables robots to process vast amounts of sensory data and act with precision in unpredictable environments, making it particularly well-suited for high-stakes applications (Gronauer and Diepold 2022). Furthermore, Multi-Agent Reinforcement Learning (MARL) has emerged as a critical approach for scenarios where robots interact not only with humans but also with other robots. MARL systems excel in situations requiring group dynamics and collaborative decision-making, enhancing coordination in tasks such as assembly lines or logistics. In these contexts, robots can autonomously decide when to take over tasks, assist human colleagues, or collaborate with other robots to prevent bottlenecks, thereby improving overall system efficiency (Zhang et al. 2021; Gronauer and Diepold 2022).

Group Dynamics and Decentralised Decision-Making

While D-HRI primarily focuses on the interactions between a single human and robot, understanding group dynamics within multi-agent robotic frameworks is crucial. This understanding introduces flexibility and robustness that enhance the efficiency of dyadic interactions. Group dynamics facilitate the distribution of responsibilities among multiple agents, leading to improved task completion times and more effective resource management (Dahiya et al. 2023). For instance, in collaborative healthcare or industrial settings, a decentralised approach enables individual robots to act autonomously based on local information, thereby preventing bottlenecks and fostering a more adaptive workflow (Jiménez et al. 2018; Huang et al. 2024).

Even in dyadic HRI scenarios, insights gained from group-based interactions can help robots manage their local tasks more effectively while still pursuing overarching collaborative goals with humans. Moreover, decentralised decision-making systems offer fault tolerance and autonomy, allowing each agent to continue functioning and making decisions even if parts of the system fail (Jiménez et al. 2018; Güzel et al. 2019). This adaptability is critical when scaling dyadic interactions to more complex multi-robot or multi-human tasks, as the knowledge acquired in decentralised multi-agent systems can directly inform the design of more sophisticated dyadic interactions.

Hybrid Decision-Making Models

Having explored various decision-making frameworks, recent research has shifted towards hybrid decision-making models that combine the strengths of different approaches to offer more comprehensive solutions. For example, integrating Reinforcement Learning (RL) with Partially Observable Markov Decision Processes (POMDPs) enables robots to learn optimal behaviours through trial and error while simultaneously managing uncertainty across varying levels of interaction (Zheng et al. 2018; Zhang et al. 2023). This hybridization proves particularly valuable in dynamic, real-world scenarios such as Human-Robot Interaction (HRI), where robots must anticipate human intentions and respond to incomplete or noisy information. By merging the adaptability of RL with the capacity of POMDPs to handle partial observability, these models empower robots to become more proactive, ultimately enhancing collaborative efficiency (Serrano et al. 2021; Xiang et al. 2021).

1.7 Safety, Ethical, and Societal Considerations in Dyadic HRI

So far, we have mapped out the stages of dyadic interaction from the perspective of a robotic framework. However, when viewed from the human perspective, concerns regarding safety and ethics become paramount, as ensuring that human-robot interactions remain safe, ethically grounded, and socially acceptable is crucial for fostering trust and cooperation between humans and robots. These systems will increasingly be deployed in sensitive environments such as healthcare, education, and public spaces. Therefore, ensuring trustworthiness, safety, and widely accepted ethical standards are incorporated into their frameworks is essential without compromising system efficiency and functionality (Valori et al. 2023; Wullenkord and Eyssel 2020). Failure to do so risks undermining public confidence and could lead to unforeseen societal consequences, particularly in areas where human well-being is directly impacted (Hamad et al. 2023; Valori et al. 2023). In Fig. 1.2, we have placed this stage prior to the"Decision-making and Action Execution" stage, reflecting the necessity for all decisions and actions taken by a robot to be informed by safety and ethical considerations.

To achieve the goals of maintaining safety, ethical standards, and societal acceptance in dyadic HRI, several key steps and considerations must be implemented.

First, **ensuring physical safety** in collaborative environments is essential, particularly as robots and humans share the same workspace (Wullenkord and Eyssel 2020). Achieving this will require advanced sensors and actuators that allow for the detection of human proximity and adjustment of robot behaviours, such as halting movement or reducing force to prevent injuries (Hamad et al. 2023; Wullenkord and Eyssel 2020; Haddadin 2015). Additionally, **failure mitigation and risk management** are vital. Robots must be designed with fail-safe mechanisms, redundancy systems, and real-time monitoring to handle potential faults and minimise risks (Hamad et al. 2023; Haddadin et al. 2011). For instance, incorporating emergency stop functionalities or self-diagnostic capabilities can reduce the likelihood of system malfunctions leading to harm. Furthermore, the concept of **soft robotics** is gaining traction, where robots are designed with compliant materials that reduce the risk of injury in case of accidental contact with humans (Wang et al. 2024).

From an **ethical standpoint**, maintaining transparency and interpretability in robot decision-making is crucial for fostering trust. Ethical guidelines must ensure that robots respect human autonomy, privacy, and dignity, particularly in sensitive sectors like healthcare and education. Robots should be programmed with ethical decision-making frameworks, ensuring their actions align with societal norms and values. Adopting principles such as those from the **ETHOS** framework can help robots prioritise ethical considerations, including fairness, privacy, and bias mitigation (Jedličková 2024; Ferrell et al. 2024). Furthermore, **regulatory considerations** must be integrated into robot design and deployment. Many regions are developing legal frameworks to govern the deployment of robots, ensuring they adhere to strict safety, ethical, and societal guidelines. Emerging frameworks, like the European Union's **AI Act**, are shaping how autonomous systems function in public spaces, dictating acceptable risk levels and outlining how robots can be held accountable for their actions (Ferrell et al. 2024; S. C. University n.d.). By embedding ethical decision-making and adhering to evolving standards, robots can be designed to operate transparently and safely, ultimately contributing to broader societal acceptance and trust.

To ensure safety, ethical standards, and societal acceptance in dyadic HRI, several comprehensive frameworks have been developed. One such programme is the **Assuring Autonomy International Programme (AAIP)** at the University of York (University of York n.d.). This programme provides guidance for the assurance of robotics and autonomous systems (RAS), addressing the technical and ethical challenges posed by their integration into society (University of York n.d.). AAIP offers tools like the **AMLAS (Assurance of Machine Learning for Autonomous Systems)** methodology, which integrates safety assurance into the development lifecycle of machine learning components used in autonomous systems. This approach ensures that RAS can be deployed safely and transparently, particularly in complex environments such as healthcare and public spaces. The AAIP framework also includes **SACE (Safety Assurance of Autonomous Systems in Complex Environments)**, which provides strategies for the safe deployment of autonomous systems in dynamic, real-world scenarios. Both AMLAS and SACE aim to foster public confidence by addressing the core technical and regulatory issues related to the safety, ethics, and transparency of autonomous systems (University of York n.d.). Integrating frameworks like AMLAS and SACE into dyadic HRI contexts would not only ensure human physical safety but also reinforce public trust in the ethical use of robotics, thereby enabling these systems to be responsibly deployed in sensitive areas such as healthcare, education, and urban environments.

1.8 Future Directions and Opportunities

Considering the technological advancements explored throughout this work, several gaps and opportunities remain for further contributions to the field of dyadic HRI. As robots become increasingly integrated into sensitive and complex environments, the future of dyadic HRI will require innovations that enhance robot autonomy and their capacity to understand and navigate human social, emotional, and cognitive states.

Upon reflection on the various stages of dyadic interaction, several unexplored avenues emerge, each offering exciting potential for advancing human-robot interaction. Below, we highlight critical opportunities for future research, structured along key stages of interaction.

1.8.1 Enhanced Perception Through Advanced Imaging and Sensing Modalities

Currently, robots primarily rely on traditional sensory modalities like vision and auditory processing when interacting with humans. However, advanced technologies such as **thermal imaging** and **hyperspectral imaging** offer new opportunities to expand robotic perception. Incorporating such technologies would allow robots to detect heat signatures or analyse the material composition of objects, which is particularly useful in high-precision tasks like medical diagnostics or hazardous material handling (Filippini et al. 2020). In dyadic HRI, this enhanced perceptual capacity could improve task efficiency and safety in environments like hospitals, eldercare facilities, and search-and-rescue operations. Beyond these applications, advanced imaging could facilitate human-robot collaboration in complex manufacturing processes, where detecting structural flaws or material inconsistencies, beyond human perception, is crucial. For example, a robot using hyperspectral imaging could identify imperfections in materials during construction, flagging potential issues before they become critical. In agriculture, hyperspectral imaging could assess crop health or detect early signs of disease, contributing to food security and sustainability. Key technological and research opportunities in this domain include:

- **Development of Multimodal Perception Frameworks**: The integration of advanced sensors such as thermal and hyperspectral imaging offers opportunities for developing multimodal perception systems. These frameworks would enable robots to merge sensory data—such as visual, thermal, and spectral inputs—enhancing their environmental understanding and decision-making capabilities, improving robot autonomy and collaboration in dyadic interactions.

- **Creation of Novel Sensing Algorithms and Machine Learning Models**: Developing machine learning models that can process and interpret advanced imaging data, such as hyperspectral and thermal signatures, represents a significant opportunity. Current vision-based models may not fully exploit the information provided by these sensors, requiring the development of new architectures that can leverage the unique features of these modalities.
- **Human-Centric Data Interpretation and Predictive Analytics**: The integration of predictive analytics with thermal and hyperspectral data offers an opportunity for enhanced understanding of human physiological states. Robots could predict stress, fatigue, or health conditions by interpreting subtle cues in these imaging outputs, leading to more intuitive and proactive robotic responses.
- **Sensor Miniaturisation and Power Efficiency**: Reducing the size and power consumption of advanced sensors is crucial for their integration into mobile robots. Current hyperspectral and thermal sensors are bulky and power-intensive, limiting their use in mobile robotics. Progress in sensor miniaturisation would significantly improve robot mobility and energy efficiency.
- **Integration with Other Sensing Modalities (e.g. Proprioception, Haptics)**: Fusing advanced imaging with proprioception and haptic feedback presents an exciting opportunity. Such integration would enable robots to combine thermal or spectral perception with tactile feedback and self-awareness, enhancing dexterity and task performance in precision-demanding environments.
- **Development of Standardised Data Processing Protocols**: Advanced imaging modalities generate vast data streams, necessitating the development of standardised protocols for processing, fusing, and interpreting this data in real time. These protocols would enhance interoperability across robotic systems and facilitate adoption across various industries.

By focusing on these technological opportunities, the field of dyadic HRI can advance towards creating more sophisticated robots capable of operating in dynamic, human-centric environments and performing complex tasks.

1.8.2 Expanding Sensory Capabilities Beyond Human Limitations

While robots are designed to complement human capabilities, there remains a significant opportunity to extend their sensory capabilities beyond traditional human perceptions. Technologies such as **enhanced electronic noses** (for detecting chemical compositions and pheromones) and **body worn wireless biosignal sensors** that monitor physiological parameters offer potential advancements. These technologies could enable robots to better understand and respond to the nuanced needs of humans in diverse environments. Key opportunities include:

- **Development of Advanced Chemical Sensing Capabilities**: Research into electronic noses, which mimic human olfactory systems to detect chemicals, offers an exciting opportunity to enhance robotic sensory abilities. Robots equipped with these sensors could analyse air quality, detect hazardous substances, or even identify human pheromones, providing new situational awareness in environments such as healthcare, disaster recovery, and industrial settings.
- **Integration of Biosignal Streaming and Health Monitoring**: The integration of wearable sensors capable of streaming real-time physiological data such as heart rate, temperature, skin pH, and perspiration levels presents an opportunity for robots to dynamically monitor human well-being. This could enable personalised care, where robots respond to changes in a human's physical condition, providing timely assistance or alerting healthcare professionals to critical changes.
- **Biorhythm and Pheromone Detection for Social Robotics**: Incorporating pheromone detection into social robots offers new possibilities for understanding human emotional states. For instance, robots in educational or therapeutic settings could interpret stress or anxiety through chemical signals, leading to more personalised and emotionally intelligent interactions.
- **Data Fusion for Enhanced Environmental Awareness**: Combining inputs from advanced olfactory sensors, thermal imaging, and biosignals could give robots a more comprehensive understanding of their environment and human partners. This multimodal sensory approach would enable robots to anticipate needs, detect hazards, and adapt their behaviour to improve collaboration.

Integrating chemical, physiological, and multisensory data pushes the boundaries of dyadic HRI. By surpassing human perceptual limitations, robots could offer more informed, precise, and empathetic responses, enhancing their effectiveness in a wide range of applications.

1.8.3 Opportunities in Predictive Modelling and Proactive Action through Physiological Monitoring

The capacity for robots to predict human actions and states based on physiological signals, such as heart rate, body temperature, and muscle tension, offers a significant future direction in dyadic HRI. The integration of real-time streaming data from human body sensors into robotic systems opens new pathways for enhancing the proactive nature of robots, allowing them to predict and react to human needs more intelligently. This is particularly relevant in applications where human well-being is critical, such as healthcare, where robots could detect early signs of stress or fatigue and adjust their behaviour accordingly. Key opportunities in this area include:

- **Development of Predictive Models for Health and Safety**: By leveraging machine learning techniques and integrating physiological data such as heart rate variability or respiratory patterns, robots could develop predictive models that anticipate when a human is in distress or at risk of health issues. For example, in healthcare, predictive analytics could enable robots to detect when a patient is becoming unwell, intervening before the situation escalates. These models can improve human-robot collaboration by making the robot's actions more intuitive and preemptive.
- **Proactive Assistance through Anticipation of Human Needs**: A major opportunity lies in enabling robots to anticipate and adapt to human needs based on physiological feedback. For instance, robots could detect a worker's physical fatigue and offer to take over tasks, or in healthcare settings, a robot could sense discomfort in a patient and adjust its actions accordingly. Proactive intervention of this nature could significantly improve collaboration in dynamic, human-centric environments.
- **Cross-Modal Integration for Holistic Understanding**: An exciting opportunity exists in combining physiological data with other sensory modalities, such as vision, olfaction, and tactile sensing, to create a holistic understanding of a human's state. This integration would enable robots to make more informed decisions, resulting in fluid, intuitive interactions that are attuned to the physical and emotional needs of the human partner.

The integration of physiological monitoring and predictive modelling offers a powerful advancement in the proactive capabilities of robots, allowing them to anticipate and respond to human needs more effectively. This has potential to improve safety, efficiency, and user satisfaction across fields such as healthcare, industrial automation, and service robotics.

1.8.4 Advanced Human-Robot Personalisation Through Learning and Cognitive Models

One of the most promising opportunities in dyadic HRI is the development of robots capable of highly personalised interactions. Imagine robots that not only respond to human commands but also learn, adapt to, and predict human preferences and behaviours, leading to more tailored emotional and behavioural support. For this vision to materialise, robots must evolve into proactive learning agents, capable of forming cognitive models of their human counterparts.

Key opportunities for personalisation in dyadic HRI include:

- **Deep Personalisation through Cognitive Models**: Future robots could transcend static interaction paradigms by using cognitive models to learn dynamically from human behaviour. This presents an exciting opportunity for harnessing deep learning techniques that track human routines, preferences, and emotional triggers. A robot could, for instance, remember how a user prefers their coffee in the morning or anticipate when they need more space during stressful moments. Such personalisation would enhance the quality of collaboration and fundamentally shift the dynamics of human-robot relationships.

- **Long-Term Interaction Data for Behavioural Predictions**: Robots that accumulate long-term interaction data could begin to predict not just a user's immediate needs, but their future actions. By recognising patterns in human behaviour, robots could foresee needs and act in advance, streamlining tasks, reducing cognitive load, and enhancing productivity in collaborative environments, such as caregiving or workplace settings.
- **Emotionally Intelligent Systems**: Emotional intelligence is a key frontier for personalisation in HRI. Robots that go beyond basic facial recognition to detect and interpret subtle emotional signals—such as frustration, joy, or motivation—can provide emotionally attuned interaction experiences. In therapeutic settings, for example, robots could sense anxiety or distress through tone of voice or body language, adjusting their behaviour to offer reassurance without needing explicit prompts. Emotionally intelligent robots have the potential to deepen human trust and foster more natural, meaningful interactions.

The integration of advanced personalization, long-term learning, and emotional intelligence presents a tremendous opportunity to push the boundaries of dyadic HRI, enabling robots to become true partners—intuitive, adaptable, and empathetic—capable of forging deeper, more meaningful connections with humans.

1.8.5 Opportunities for Enhancing Dyadic HRI Through Advanced Spatial Awareness

The integration of cutting-edge technologies such as proxemics, environmental semantic understanding, spatial perspective-taking, and Digital Twin technology presents significant opportunities for enhancing spatial awareness in dyadic HRI. Through these frameworks, robots can develop a richer understanding of shared spaces, navigate more safely, and improve their collaborative efficiency with human partners. This leads to the possibility of more autonomous and intuitive robots that dynamically adapt to the fluid nature of human environments.

Among the most transformative technologies in this regard is **Digital Twins**, which provide real-time monitoring and adaptive modelling of environments. By continuously updating a virtual replica of the robot's surroundings, Digital Twins allow robots to anticipate human movements, respond to environmental changes, and optimise task execution without the need for constant human supervision. This technology has significant potential in personalising and contextualising interactions in settings such as hospitals, factories, and public spaces, where safety, efficiency, and task relevance are paramount.

Additionally, advancements in **semantic understanding** offer robots the capability to navigate and comprehend unstructured environments, such as homes or eldercare facilities. By understanding the relationships between objects and spaces, robots can meaningfully assist humans, reducing errors and enhancing collaboration.

Further opportunities arise in the development of **proxemics** and **spatial perspective-taking**. By respecting personal space and adopting the human perspective, robots could foster more natural, fluid interactions in both professional and everyday settings, transforming the way humans and robots coexist.

By leveraging these advancements in spatial awareness, dyadic HRI could witness the emergence of robots that not only excel in technical tasks but also navigate the complexities of human-centred environments more effectively. Ultimately, these advancements could improve both performance and acceptance across industries such as healthcare, manufacturing, and public services.

1.9 Conclusion

This paper has examined the evolving landscape of dyadic Human-Robot Interaction, with a particular focus on the technological advancements shaping its future. By exploring the integration of key technologies such as Digital Twins, Generative Artificial Intelligence, and computer vision, we have highlighted the potential for creating robots that are more adaptive, responsive, and context-aware. These technologies, when combined, provide a foundation for robots to engage in complex interactions with humans, addressing not only the technical aspects of collaboration but also its social and emotional dimensions.

The role of Digital Twins in creating real-time, virtual replicas of physical environments offers exciting opportunities for dynamic interaction and decision-making in uncertain or changing contexts. Moreover, the synergy between Digital Twins and GenAI facilitates predictive modelling and synthetic data generation, enabling robots to continuously learn and adapt to the environments in which they operate. These developments promise to enhance robot autonomy, making them more capable of anticipating human needs and improving their integration across industries such as healthcare, manufacturing, and public services.

We have also identified critical challenges and opportunities for future research, including the enhancement of spatial awareness, the advancement of multimodal perception, and the need to address the ethical, societal, and regulatory implications of advanced HRI systems. As robots become increasingly integrated into human environments, it is crucial to ensure that these interactions remain ethical, safe, and human-centred.

In summary, the future of dyadic HRI will be shaped by the convergence of these cutting-edge technologies with human-centred design principles. By continuing to advance robot perception, interaction capabilities, and adaptability, robots will not only assist humans in a wide range of tasks but also foster deeper, more intuitive collaborations that align with the goals of Industry 5.0.

References

A review of verbal and non-verbal human-robot interactive communication. Rob Autonom Syst. https://doi.org/10.1016/j.robot.2014.09.031. Accessed 24 Sep 2024

Abrini M, Auvray M, Chetouani M (2023) 'Humans' spatial perspective-taking when interacting with a robotic arm. In: 2023 32nd IEEE international conference on robot and human interactive communication (RO-MAN), pp 533–540. https://doi.org/10.1109/RO-MAN57019.2023.10309579

Adebayo S, Dessing JC, McLoone S (2025) SLYKLatent: a learning framework for gaze estimation using deep facial feature learning. IEEE Trans Hum-Mach Syst 55(3):333–346. https://doi.org/10.1109/THMS.2025.3553404

Adebayo S, McLoone S, Dessing JC (2022) Hand-eye-object tracking for human intention inference. IFAC-PapersOnLine 55(12):174–179. https://doi.org/10.1016/j.ifacol.2022.07.627

Adebayo S, McLoone S, Dessing JC (2024) QUB-PHEO: a visual-based dyadic multi-view dataset for intention inference in collaborative assembly. IEEE Access 12:157050–157066. https://doi.org/10.1109/ACCESS.2024.3485162

Admoni H, Scassellati B (2017) Social eye gaze in human-robot interaction: a review. J Hum Robot Interact 6(1):25–63. https://doi.org/10.5898/JHRI.6.1.Admoni

Ahmad M, Mubin O, Orlando J (2017) A systematic review of adaptivity in human-robot interaction. Multimodal Technol Interact 1(3) (Art. no. 3). https://doi.org/10.3390/mti1030014

Ahmadzadeh H, Masehian E (2015) Modular robotic systems: methods and algorithms for abstraction, planning, control, and synchronization. Artif Intell 223:27–64. https://doi.org/10.1016/j.artint.2015.02.004

Asad U, Khan M, Khalid A, Lughmani WA (2023) Human-centric digital twins in industry: a comprehensive review of enabling technologies and implementation strategies. Sensors 23(8):3938. https://doi.org/10.3390/s23083938

Baker C, Saxe R, Tenenbaum J (2011) Bayesian theory of mind: modeling joint belief-desire attribution. In: Proceedings of the annual meeting of the cognitive science society, vol 33. https://escholarship.org/uc/item/5rk7z59q

Baltrušaitis T, Robinson P, Morency L-P (2016) OpenFace: an open source facial behavior analysis toolkit. In: 2016 IEEE winter conference on applications of computer vision (WACV), pp 1–10. https://doi.org/10.1109/WACV.2016.7477553

Baratta A, Cimino A, Longo F, Nicoletti L (2023) Digital twin for human-robot collaboration enhancement in manufacturing systems: literature review and direction for future developments. Comput Indus Eng 109764. https://doi.org/10.1016/j.cie.2023.109764

Barresi G, Pacchierotti C, Laffranchi M, De Michieli L (2022) Beyond digital twins: phygital twins for neuroergonomics in human-robot interaction. Front Neurorobot 16:913605. https://doi.org/10.3389/fnbot.2022.913605

Baylor AL (2011) The design of motivational agents and avatars. Educ Tech Res Dev 59(2):291–300. https://doi.org/10.1007/s11423-011-9196-3

Beck A, Hiolle A, Mazel A, Cañamero L (2010) Interpretation of emotional body language displayed by robots. In: Proceedings of the 3rd international workshop on affective interaction in natural environments, in AFFINE'10. Association for Computing Machinery, New York, NY, pp 37–42. https://doi.org/10.1145/1877826.1877837

Belardinelli A (2024) Gaze-based intention estimation: principles, methodologies, and applications in HRI. J Hum Robot Interact 13(3):31:1–31:30.https://doi.org/10.1145/3656376

Bera A, Randhavane T, Prinja R, Manocha D (2017) SocioSense: robot navigation amongst pedestrians with social and psychological constraints. In: 2017 IEEE/RSJ international conference on intelligent robots and systems (IROS), pp 7018–7025. https://doi.org/10.1109/IROS.2017.8206628

Bremers A, Pabst A, Parreira MT, Ju W (2024) Using social cues to recognize task failures for HRI: overview, state-of-the-art, and future directions. arXiv: arXiv:2301.11972. Accessed 24 Sep 2024

Canny J (1986) A computational approach to edge detection. In: IEEE transactions on pattern analysis and machine intelligence, vol PAMI-8, no 6, pp 679-698. https://doi.org/10.1109/TPAMI.1986.4767851

Cao Z, Hidalgo G, Simon T, Wei S-E, Sheikh Y (2021) OpenPose: realtime multi-Person 2D pose estimation using part affinity fields. IEEE Trans Patt Anal Mach Intell 43(01):172–186

Chaminade T, Okka MM (2013) Comparing the effect of humanoid and human face for the spatial orientation of attention. Front Neurorobot 7. https://doi.org/10.3389/fnbot.2013.00012

Chen M, Hsu D, Lee WS (2019) Guided exploration of human intentions for human-robot interaction. arXiv: arXiv:1812.04728. Accessed 19 Sep 2024

Chen L, Liu Z, Wu M, Hirota K, Pedrycz W (2021) Multimodal emotion recognition and intention understanding in human-robot interaction. In:Wu M, Pedrycz W, Chen L (eds) Developments in advanced control and intelligent automation for complex systems. Springer International Publishing, Cham, pp 255–288. https://doi.org/10.1007/978-3-030-62147-6_10

Chita-Tegmark M, Scheutz M (2021) Assistive robots for the social management of health: a framework for robot design and human-robot interaction research. Int J Soc Rob 13(2):197–217. https://doi.org/10.1007/s12369-020-00634-z

Chung S, Lim J, Noh KJ, Kim G, Jeong H (2019) Sensor data acquisition and multimodal sensor fusion for human activity recognition using deep learning. Sensors 19(7) (Art. no. 7). https://doi.org/10.3390/s19071716

Churamani N, Barros P, Gunes H, Wermter S (2022) Affect-driven learning of robot behaviour for collaborative human-robot interactions. Front Robot AI 9. https://doi.org/10.3389/frobt.2022.717193

Clabaugh C et al (2019) Long-term personalization of an in-home socially assistive robot for children with autism spectrum disorders. Front Robot AI 6. https://doi.org/10.3389/frobt.2019.00110

Collins EC, Prescott TJ, Mitchinson B (2015) Saying it with light: a pilot study of affective communication using the MIRO robot. In: Wilson SP, Verschure PFMJ, Mura A, Prescott TJ, (eds) Biomimetic and biohybrid systems. Springer International Publishing, Cham, pp 243–255. https://doi.org/10.1007/978-3-319-22979-9_25

'CSISS Classics—Edward T. Hall (1966) Proxemic theory

Dahiya A, Aroyo AM, Dautenhahn K, Smith SL (2023) A survey of multi-agent human-robot interaction systems. Rob Autonom Syst 161:104335. https://doi.org/10.1016/j.robot.2022.104335

David D, Thérouanne P, Milhabet I (2022) The acceptability of social robots: a scoping review of the recent literature. Comput Human Behav 137:107419. https://doi.org/10.1016/j.chb.2022.107419

Devlin J, Chang M-W, Lee K, Toutanova K (2019) BERT: pre-training of deep bidirectional transformers for language understanding. In: Burstein J, Doran C, Solorio T (eds) Proceedings of the 2019 conference of the North American chapter of the association for computational linguistics: human language technologies, vol 1 (long and short papers). Association for Computational Linguistics, Minneapolis, MN, pp 4171–4186. https://doi.org/10.18653/v1/N19-1423

Doncieux S, Chatila R, Straube S, Kirchner F (2022) Human-centered AI and robotics. AI Perspect 4(1):1. https://doi.org/10.1186/s42467-021-00014-x

Duan J, Zhuang L, Zhang Q, Zhou Y, Qin J (2024) Multimodal perception-fusion-control and human-robot collaboration in manufacturing: a review. Int J Adv Manuf Technol 132(3):1071–1093. https://doi.org/10.1007/s00170-024-13385-2

Ferrell OC, Harrison DE, Ferrell LK, Ajjan H, Hochstein BW (2024) A theoretical framework to guide AI ethical decision making. AMS Rev 14(1):53–67. https://doi.org/10.1007/s13162-024-00275-9

Filippini C, Perpetuini D, Cardone D, Chiarelli AM, Merla A (2020) Thermal infrared imaging-based affective computing and its application to facilitate human robot interaction: a review. Appl Sci 10(8) (Art. no. 8). https://doi.org/10.3390/app10082924

Fiore SM, Wiltshire TJ, Lobato EJ, Jentsch FG, Huang WH, Axelrod B (2013) Toward understanding social cues and signals in human-robot interaction: effects of robot gaze and proxemic behavior. Front Psychol 4. https://doi.org/10.3389/fpsyg.2013.00859

Frederiksen MR (2021) Toward context-aware, affective, and impactful social robots. In: ITU-DS. IT-Universitetet i København

Fudenberg D, Newey WK, Strack P, Strzalecki T (2020) Testing the drift-diffusion model. Proc Natl Acad Sci USA 117(52):33141–33148. https://doi.org/10.1073/pnas.2011446117

Gallala A, Kumar AA, Hichri B, Plapper P (2022) Digital twin for human–robot interactions by means of industry 4.0 enabling technologies. Sensors 22(13) (Art. no. 13). https://doi.org/10.3390/s22134950

Gallala A, Kumar AA, Hichri B, Plapper P (2022) Digital twin for human–robot interactions by means of industry 4.0 enabling technologies. Sensors 22(13):4950. https://doi.org/10.3390/s22134950

Ghazali AS, Ham J, Barakova E, Markopoulos P (2019) Assessing the effect of persuasive robots interactive social cues on users' psychological reactance, liking, trusting beliefs and compliance. Adv Rob 33(7–8):325–337. https://doi.org/10.1080/01691864.2019.1589570

Ginerica C, Zaha M, Floroian L, Cojocaru D, Grigorescu S (2024) A vision dynamics learning approach to robotic navigation in unstructured environments. Robotics 13(1) (Art. no. 1). https://doi.org/10.3390/robotics13010015

Gong T et al (2024) Multimodal fusion and human-robot interaction control of an intelligent robot. Front Bioeng Biotechnol 11. https://doi.org/10.3389/fbioe.2023.1310247

Grichkovtsova I, Morel M, Lacheret A (2012) The role of voice quality and prosodic contour in affective speech perception. Speech Commun 54(3):414–429. https://doi.org/10.1016/j.specom.2011.10.005

Gronauer S, Diepold K (2022) Multi-agent deep reinforcement learning: a survey. Artif Intell Rev 55(2):895–943. https://doi.org/10.1007/s10462-021-09996-w

Guastella DC, Muscato G (2021) Learning-based methods of perception and navigation for ground vehicles in unstructured environments: a review. Sensors 21(1) (Art. no. 1). https://doi.org/10.3390/s21010073

Gunes H, Celiktutan O, Sariyanidi E (2019) Live human-robot interactive public demonstrations with automatic emotion and personality prediction. Philos Trans R Soc B Biol Sci 374(1771):20180026. https://doi.org/10.1098/rstb.2018.0026

Güzel MS, Ajabshir VB, Nattharith P, Gezer EC, Can S (2019) A novel framework for multi-agent systems using a decentralized strategy. Robotica 37(4):691–707. https://doi.org/10.1017/S0263574718001261

Haddadin S (2015) Physical safety in robotics. In: Drechsler R, Kühne U (eds) Formal modeling and verification of cyber-physical systems: 1st international summer school on methods and tools for the design of digital systems, Bremen, Germany. Springer Fachmedien, Wiesbaden, pp 249–271. https://doi.org/10.1007/978-3-658-09994-7_9

Haddadin S, Albu-Schäffer A, Hirzinger G (2011) Safe physical human-robot interaction: measurements, analysis and new insights. In: Kaneko M, Nakamura Y (eds) Robotics research. Springer, Berlin, Heidelberg, pp 395–407. https://doi.org/10.1007/978-3-642-14743-2_33

Hamad M, Nertinger S, Kirschner RJ, Figueredo L, Naceri A, Haddadin S (2023) A concise overview of safety aspects in human-robot interaction. In: Piazza C, Capsi-Morales P, Figueredo L, Keppler M, Schütze H (eds) Human-friendly robotics 2023. Springer Nature Switzerland, Cham, pp 1–18. https://doi.org/10.1007/978-3-031-55000-3_1

Hanheide M, Peters A, Bellotto N (2012) Analysis of human-robot spatial behaviour applying a qualitative trajectory calculus. In: 2012 IEEE RO-MAN: the 21st IEEE international symposium on robot and human interactive communication, pp 689–694. https://doi.org/10.1109/ROMAN.2012.6343831

Hannun AY et al. (2024) Deep speech: scaling up end-to-end speech recognition. https://www.semanticscholar.org/paper/Deep-Speech%3A-Scaling-up-end-to-end-speech-Hannun-Case/24741d280869ad9c60321f5ab6e5f01b7852507d. Accessed 4 Oct 2024

Harris CG, Stephens MJ (1988) A combined corner and edge detector. In: Alvey Vision Conference
Hayhoe MM, Shrivastava A, Mruczek R, Pelz JB (2003) Visual memory and motor planning in a natural task. J Vis 3(1):6. https://doi.org/10.1167/3.1.6
Henschel A, Laban G, Cross ES (2021) What makes a robot social? A review of social robots from science fiction to a home or hospital near you. Curr Robot Rep 2(1):9–19. https://doi.org/10.1007/s43154-020-00035-0
Hoffman G, Bhattacharjee T, Nikolaidis S (2024) Inferring human intent and predicting human action in human-robot collaboration. Ann Rev Control Rob Autonom Syst 7(1):73–95. https://doi.org/10.1146/annurev-control-071223-105834
Hoffman G, Breazeal C (2004) Collaboration in human-robot teams. In: AIAA 1st intelligent systems technical conference. American Institute of Aeronautics and Astronautics, Chicago, IL. https://doi.org/10.2514/6.2004-6434
Hoffman G, Zhao X (2020) A primer for conducting experiments in human–robot interaction. J Hum Robot Interact 10(1):6:1–6:31. https://doi.org/10.1145/3412374
Hong A, Tsuboi Y, Nejat G, Benhabib B (2017) Multimodal affect recognition for assistive human-robot interactions. Presented at the design of medical devices conference. American Society of Mechanical Engineers Digital Collection. https://doi.org/10.1115/DMD2017-3332
Huang KA, Choudhary HK, Kuo PC (2024) Artificial intelligent agent architecture and clinical decision-making in the healthcare sector. Cureus 16(7):e64115. https://doi.org/10.7759/cureus.64115
Huber A, Lammer L, Weiss A, Vincze M (2014) Designing adaptive roles for socially assistive robots: a new method to reduce technological determinism and role stereotypes. J Hum Robot Interact 3(2):100–115. https://doi.org/10.5898/JHRI.3.2.Huber
Huettenrauch H, Eklundh KS, Green A, Topp EA (2006) Investigating spatial relationships in human-robot interaction. In: 2006 IEEE/RSJ international conference on intelligent robots and systems, pp 5052–5059. https://doi.org/10.1109/IROS.2006.282535
Human-machine collaborative decision-making: an evolutionary roadmap based on cognitive intelligence. Int J Social Rob. Accessed 5 Oct 2024. https://link.springer.com/article/10.1007/s12369-023-01020-1
Hussain S, Saeed K, Baimagambetov A, Rab S, Saad M (2024) Advancements in gesture recognition techniques and machine learning for enhanced human-robot interaction: a comprehensive review. arXiv:2409.06503. https://doi.org/10.48550/arXiv.2409.06503
Inagaki Y, Sugie H, Aisu H, Ono S, Unemi T (1995) Behavior-based intention inference for intelligent robots cooperating with human. In: Proceedings of 1995 IEEE international conference on fuzzy systems, vol 3, pp 1695–1700. https://doi.org/10.1109/FUZZY.1995.409904
Industry 5.0—European Commission. Accessed 19 Sep 2024. https://research-and-innovation.ec.europa.eu/research-area/industrial-research-and-innovation/industry-50_en
Iovino M, Förster J, Falco P, Chung JJ, Siegwart R, Smith C (2023) On the programming effort required to generate behavior trees and finite state machines for robotic applications. In: 2023 IEEE international conference on robotics and automation (ICRA), pp 5807–5813. https://doi.org/10.1109/ICRA48891.2023.10160972
Iovino M, Förster J, Falco P, Chung J, Siegwart R, Smith C (2024) Comparison between behavior trees and finite state machines., arXiv: arXiv:2405.16137. Accessed 5 Oct 2024
Itti L, Baldi P (2009) Bayesian surprise attracts human attention. Vis Res 49(10):1295–1306. https://doi.org/10.1016/j.visres.2008.09.007
Jahanmahin R, Masoud S, Rickli J, Djuric A (2022) Human-robot interactions in manufacturing: a survey of human behavior modeling. Rob Comput Integr Manuf 78:102404. https://doi.org/10.1016/j.rcim.2022.102404
Jahshan C, Wynn JK, Green MF (2013) Relationship between auditory processing and affective prosody in schizophrenia. Schizophrenia Res 143(2):348–353. https://doi.org/10.1016/j.schres.2012.11.025
Jain S, Argall B (2019) Probabilistic human intent recognition for shared autonomy in assistive robotics. J Hum Robot Interact 9(1):2:1–2:23. https://doi.org/10.1145/3359614

Jarrassé N, Sanguineti V, Burdet E (2014) Slaves no longer: review on role assignment for human-robot joint motor action. Adap Behav 22(1):70–82. https://doi.org/10.1177/1059712313481044
Jedličková A (2024) Ethical approaches in designing autonomous and intelligent systems: a comprehensive survey towards responsible development. AI Soc. https://doi.org/10.1007/s00146-024-02040-9
Jiménez AC, García-Díaz V, Bolaños S (2018) A decentralized framework for multi-agent robotic systems. Sensors 18(2) (Art. no. 2). https://doi.org/10.3390/s18020417
Jirak D, Sciutti A, Barros P, Rea F (2022) I am only happy when there is light: the impact of environmental changes on affective facial expressions recognition. https://doi.org/10.48550/arXiv.2210.17421
Jobanputra C, Bavishi J, Doshi N (2019) Human activity recognition: a survey. Proc Comput Sci 155:698–703. https://doi.org/10.1016/j.procs.2019.08.100
Johansson RS, Westling G, Bäckström A, Flanagan JR (2001) Eye-hand coordination in object manipulation. J Neurosci 21(17):6917–6932. https://doi.org/10.1523/JNEUROSCI.21-17-06917.2001
Koppula HS, Saxena A (2014) Physically grounded spatio-temporal object affordances. In:Fleet D, Pajdla T, Schiele B, Tuytelaars T (eds) Computer vision—ECCV 2014, in lecture notes in computer science, vol 8691. Springer International Publishing, Cham, pp 831–847. https://doi.org/10.1007/978-3-319-10578-9_54
Kruse T, Pandey AK, Alami R, Kirsch A (2013) Human-aware robot navigation: a survey. Rob Autonom Syst 61(12):1726–1743. https://doi.org/10.1016/j.robot.2013.05.007
Kulke L, Feyerabend D, Schacht A (2020) A comparison of the affectiva iMotions facial expression analysis software with EMG for identifying facial expressions of emotion. Front Psychol 11. https://doi.org/10.3389/fpsyg.2020.00329
Kuo I-H, Jayawardena C, Broadbent E, MacDonald BA (2011) Multidisciplinary design approach for implementation of interactive services. Int J Soc Rob 3(4):443–456. https://doi.org/10.1007/s12369-011-0115-x
Kwak SS (2014) The impact of the robot appearance types on social interaction with a robot and service evaluation of a robot. ADR. https://doi.org/10.15187/adr.2014.05.110.2.81
Kwak SS (2014) The impact of the robot appearance types on social interaction with a robot and service evaluation of a robot. ADR. https://doi.org/10.15187/adr.2014.05.110.2.81
Kwasny SC, Faisal KA (1990) Overcoming limitations of rule-based systems: an example of a hybrid deterministic parser. In: Dorffner G (ed) Konnektionismus in artificial intelligence und Kognitionsforschung. Springer, Berlin, Heidelberg, pp 48–57. https://doi.org/10.1007/978-3-642-76070-9_5
Land MF (2009) Vision, eye movements, and natural behavior. Vis Neurosci 26(1):51–62. https://doi.org/10.1017/S0952523808080899
Lauri M, Hsu D, Pajarinen J (2023) Partially observable Markov decision processes in robotics: a survey. IEEE Trans Rob 39(1):21–40. https://doi.org/10.1109/TRO.2022.3200138
Lawrence N (2003) Gaussian process latent variable models for visualisation of high dimensional data. In: Advances in neural information processing systems. vol 16, MIT Press
Lemaignan S, Warnier M, Sisbot EA, Clodic A, Alami R (2017) Artificial cognition for social human-robot interaction: an implementation. Artif Intell 247:45–69. https://doi.org/10.1016/j.artint.2016.07.002
Leng J, Sha W, Wang B, Zheng P, Zhuang C, Liu Q, Wuest T, Mourtzis D, Wang L (2022) Industry 5.0: prospect and retrospect. J Manuf Syst 65:279–295
Leusmann J, Wang C, Mayer S (2024) Comparing rule-based and LLM-based methods to enable active robot assistant conversations. In: CUI@CHI 2024: building trust in CUIs—from design to deployment 05-11
Li AX, Florendo M, Miller LE, Ishiguro H, Saygin AP (2015) Robot form and motion influences social attention. In: Proceedings of the tenth annual ACM/IEEE international conference on human-robot interaction, in HRI '15. Association for Computing Machinery, New York, NY, pp 43–50. https://doi.org/10.1145/2696454.2696478

Li J, Ngai G, Leong HV, Chan SCF (2016) Multimodal human attention detection for reading from facial expression, eye gaze, and mouse dynamics. SIGAPP Appl Comput Rev 16(3):37–49. https://doi.org/10.1145/3015297.3015301

Liao Z, Cai Y (2024) AR-enhanced digital twin for human-robot interaction in manufacturing systems. Energ Ecol Environ. https://doi.org/10.1007/s40974-024-00327-7

Lima V, Zanini MT, Reis Irigaray HA (2022) Non-dyadic human-robot interactions and online brand communities. Mark Intell Plan 40(6):724–737. https://doi.org/10.1108/MIP-02-2022-0059

Lindsay A, Ramírez-Duque AA, Petrick RPA, Foster ME (2024) A socially assistive robot using automated planning in a paediatric clinical setting. In: Proceedings of the 2024 international symposium on technological advances in human-robot interaction. ACM, Boulder, CO, pp 47–55. https://doi.org/10.1145/3648536.3648542

Liu W, Bao Q, Sun Y, Mei T (2022) Recent advances of monocular 2D and 3D human pose estimation: a deep learning perspective. ACM Comput Surv 55(4):80:1–80:41. https://doi.org/10.1145/3524497

Liu L, Liu Y, Gao X-Z (2021) Impacts of human robot proxemics on human concentration-training games with humanoid robots. Healthcare 9(7) (Art. no. 7). https://doi.org/10.3390/healthcare9070894

Louie W-YG, Vaquero T, Nejat G, Beck JC (2014) An autonomous assistive robot for planning, scheduling and facilitating multi-user activities. In: 2014 IEEE international conference on robotics and automation (ICRA), pp 5292–5298. https://doi.org/10.1109/ICRA.2014.6907637

Lu S-R, Lo J-H, Hong Y-T, Huang H-P (2024) Implementation of engagement detection for human–robot interaction in complex environments. Sensors 24(11) (Art. no. 11). https://doi.org/10.3390/s24113311

Lu W, Chen J, Fu Y, Pan Y, Ghansah FA (2023) Digital twin-enabled human-robot collaborative teaming towards sustainable and healthy built environments. J Clean Prod 412:137412. https://doi.org/10.1016/j.jclepro.2023.137412

Lu W, Chen J, Fu Y, Pan Y, Ghansah FA (2023) Digital twin-enabled human-robot collaborative teaming towards sustainable and healthy built environments. J Clean Prod 412:137412. https://doi.org/10.1016/j.jclepro.2023.137412

Luo Y, Ye J, Adams RB, Li J, Newman MG, Wang JZ (2020) ARBEE: towards automated recognition of bodily expression of emotion in the wild. Int J Comput Vis 128(1):1–25. https://doi.org/10.1007/s11263-019-01215-y

Lutfi SBL (2013) User-centric need-driven affect modeling for spoken conversational agents: design and evaluation. https://doi.org/10.20868/upm.thesis.15839

Machine learning for human emotion recognition: a comprehensive review. Neural Comput Appl. Accessed 5 Oct 2024. https://link.springer.com/article/10.1007/s00521-024-09426-2

Malaiya RK (2024) Fitting a drift–diffusion item response theory model to complex cognition response times. In: Wiberg M, Kim J-S, Hwang H, Wu H, Sweet T (eds) Quantitative psychology. Springer Nature Switzerland, Cham, pp 199–209.https://doi.org/10.1007/978-3-031-55548-0_19

Malik AA, Brem A (2021) Digital twins for collaborative robots: a case study in human-robot interaction. Rob Comput Integr Manuf 68:102092. https://doi.org/10.1016/j.rcim.2020.102092

McColl D, Zhang Z, Nejat G (2011) Human body pose interpretation and classification for social human-robot interaction. Int J Soc Rob 3(3):313–332. https://doi.org/10.1007/s12369-011-0099-6

McColl D, Hong A, Hatakeyama N, Nejat G, Benhabib B (2016) A survey of autonomous human affect detection methods for social robots engaged in natural HRI. J Intell Robot Syst 82(1):101–133. https://doi.org/10.1007/s10846-015-0259-2

McColl D, Nejat G (2012) Affect detection from body language during social HRI. In: IEEE RO-MAN: the 21st IEEE international symposium on robot and human interactive communication. Paris, pp 1013–1018. https://doi.org/10.1109/ROMAN.2012.6343882 (Keywords: Robot sensing systems; Head; Ellipsoids; Skin ; Humans; Estimation)

Mead R, Matarić MJ (2017) Autonomous human-robot proxemics: socially aware navigation based on interaction potential'. Autonom Robot 41(5):1189–1201. https://doi.org/10.1007/s10514-016-9572-2
Mead R, and Matarić MJ (2016) Perceptual models of human-robot proxemics. In: Hsieh MA, Khatib O, Kumar V, (eds) Experimental robotics: the 14th international symposium on experimental robotics. Springer International Publishing, Cham, pp 261–276. https://doi.org/10.1007/978-3-319-23778-7_18
Mead R, Atrash A, Matarić MJ (2011) Proxemic feature recognition for interactive robots: automating metrics from the social sciences. In: Mutlu B, Bartneck C, Ham J, Evers V, Kanda T (eds) Social robotics. Springer, Berlin, Heidelberg, pp 52–61.https://doi.org/10.1007/978-3-642-25504-5_6
MediaPipe. A framework for perceiving and processing reality. http://research.google/pubs/mediapipe-a-framework-for-perceiving-and-processing-reality/. Accessed 5 Oct 2024
Miller ME, Spatz E (2022) A unified view of a human digital twin. Hum Intell Syst Integr 4(1):23–33. https://doi.org/10.1007/s42454-022-00041-x
Mohammed SN, Hassan AKA (2021) A survey on emotion recognition for human robot interaction. CIT J Comput Inform Technol 28(2) (Art. no. 2)
Mörtl A, Lawitzky M, Kucukyilmaz A, Sezgin M, Basdogan C, Hirche S (2012) The role of roles: physical cooperation between humans and robots. Int J Rob Res 31(13):1656–1674. https://doi.org/10.1177/0278364912455366
Myers CE, Interian A, Moustafa AA (2022) A practical introduction to using the drift diffusion model of decision-making in cognitive psychology, neuroscience, and health sciences. Front Psychol 13. https://doi.org/10.3389/fpsyg.2022.1039172
Nahavandi S (2019) Industry 5.0—a human-centric solution. Sustainability 11(16) (Art. no. 16). https://doi.org/10.3390/su11164371
Neuteboom SY, de Graaf MMA (2021) People's perceptions of gendered robots performing gender stereotypical tasks. In: Li H, Ge SS, Wu Y, Wykowska A, He H, Liu X, Li D, Perez-Osorio J (eds) Social robotics. Springer International Publishing, Cham, pp 24–35. https://doi.org/10.1007/978-3-030-90525-5_3
Nguyen T, Nichols E, Gomez R (2023) A study on social robot behavior in group conversation. arXiv: arXiv:2312.12473. Accessed 26 Sep 2024
Nolfi S (2022) Progress and challenges in adaptive robotics. Front Robot AI 9. https://doi.org/10.3389/frobt.2022.1020462
Obaid M, Häring M, Kistler F, Bühling R, André E (2012) User-defined body gestures for navigational control of a humanoid robot. In: Ge SS, Khatib O, Cabibihan J-J, Simmons R, Williams M-A (eds) Social robotics. Springer, Berlin, Heidelberg, pp 367–377. https://doi.org/10.1007/978-3-642-34103-8_37
Obaigbena A et al (2024) AI and human-robot interaction: a review of recent advances and challenges. GSC Adv Res Rev 18(2):321–330. https://doi.org/10.30574/gscarr.2024.18.2.0070
Onnasch L, Schweidler P, Schmidt H (2023) The potential of robot eyes as predictive cues in HRI-an eye-tracking study. Front Rob AI 10:1178433. https://doi.org/10.3389/frobt.2023.1178433
O'Sullivan JA et al (2015) Attentional selection in a cocktail party environment can be decoded from single-trial EEG. Cerebral Cortex 25(7):1697–1706. https://doi.org/10.1093/cercor/bht355
Pacherie E (2008) The phenomenology of action: a conceptual framework. Cognition 107(1):179–217. https://doi.org/10.1016/j.cognition.2007.09.003
Pathirana P, Senarath S, Meedeniya D, Jayarathna S (2022) Eye gaze estimation: a survey on deep learning-based approaches. Exp Syst Appl 199:116894. https://doi.org/10.1016/j.eswa.2022.116894
Paulmann S, Weinstein N, Zougkou K (2019) Now listen to this! Evidence from a cross-spliced experimental design contrasting pressuring and supportive communications. Neuropsychologia 124:192–201. https://doi.org/10.1016/j.neuropsychologia.2018.12.011
Pelz JB, Canosa R (2001) Oculomotor behavior and perceptual strategies in complex tasks. Vis Res 41(25):3587–3596. https://doi.org/10.1016/S0042-6989(01)00245-0
Pose—Ultralytics YOLO Docs. Accessed 5 Oct 2024. https://docs.ultralytics.com/tasks/pose/

Powers A, Kiesler S, Fussell S, Torrey C (2007) Comparing a computer agent with a humanoid robot. In: Proceedings of the ACM/IEEE international conference on Human-robot interaction. In: HRI '07. Association for Computing Machinery, New York, NY, pp 145–152. https://doi.org/10.1145/1228716.1228736

Puig-Pey A et al (2022) A new methodological approach to analyze human roles in human-robot interaction scenarios. In: 2022 IEEE international conference on advanced robotics and its social impacts (ARSO), pp 1–6. https://doi.org/10.1109/ARSO54254.2022.9802978

Qi J, Ma L, Cui Z, Yu Y (2024) Computer vision-based hand gesture recognition for human-robot interaction: a review. Compl Intell Syst 10(1):1581–1606. https://doi.org/10.1007/s40747-023-01173-6

Radford A, Narasimhan K, Salimans T, Sutskever I (2018) Improving language understanding by generative pre-training

Ramasubramanian AK, Mathew R, Kelly M, Hargaden V, Papakostas N (2022) Digital twin for human–robot collaboration in manufacturing: review and outlook. Appl Sci 12(10) (Art. no. 10). https://doi.org/10.3390/app12104811

Redmon J, Divvala S, Girshick R, Farhadi A (2016) You only look once: unified, real-time object detection. In: 2016 IEEE conference on computer vision and pattern recognition (CVPR). IEEE, Las Vegas, NV, pp 779–788. https://doi.org/10.1109/CVPR.2016.91

Ren S, He K, Girshick R, Sun J (2017) Faster R-CNN: towards real-time object detection with region proposal networks. IEEE Trans Patt Anal Mach Intell 39(06):1137–1149

Rios-Martinez J, Spalanzani A, Laugier C (2015) From proxemics theory to socially-aware navigation: a survey. Int J Soc Rob 7(2):137–153. https://doi.org/10.1007/s12369-014-0251-1

Rizzolatti G, Riggio L, Dascola I, Umiltá C (1987) Reorienting attention across the horizontal and vertical meridians: evidence in favor of a premotor theory of attention. Neuropsychologia 25(1-A)31–40. https://doi.org/10.1016/0028-3932(87)90041-8

Roychoudhury A, Khorshidi S, Agrawal S, Bennewitz M (2023) Perception for humanoid robots. Curr Robot Rep 4(4):127–140. https://doi.org/10.1007/s43154-023-00107-x

Rudovic O, Park HW, Busche J, Schuller B, Breazeal C, Picard RW (2019) Personalized estimation of engagement from videos using active learning with deep reinforcement learning. In: 2019 IEEE/CVF conference on computer vision and pattern recognition workshops (CVPRW), pp 217–226. https://doi.org/10.1109/CVPRW.2019.00031

Rusu RB, Marton ZC, Blodow N, Dolha M, Beetz M (2008) Towards 3D point cloud based object maps for household environments. Rob Autonom Syst 56(11):927–941. https://doi.org/10.1016/j.robot.2008.08.005

Sajko G, Babič J (2024) Achieving human-inspired drift diffusion consensus in swarm robotics. In: Hamann H, Dorigo M, Pérez Cáceres L, Reina A, Kuckling J, Kaiser TK, Soorati M, Hasselmann K, Buss E (eds) Swarm intelligence. Springer Nature Switzerland, Cham, pp 29–41. https://doi.org/10.1007/978-3-031-70932-6_3

Sandoval EB, Brandstatter J, Yalcin U, Bartneck C (2021) Robot likeability and reciprocity in human robot interaction: using ultimatum game to determinate reciprocal likeable robot strategies. Int J Soc Robo 13(4):851–862. https://doi.org/10.1007/s12369-020-00658-5

Sarker IH, Janicke H, Ferrag MA, Abuadbba A (2024) Multi-aspect rule-based AI: methods, taxonomy, challenges and directions towards automation, intelligence and transparent cybersecurity modeling for critical infrastructures. Internet Things 25:101110. https://doi.org/10.1016/j.iot.2024.101110

Schroff F, Kalenichenko D, Philbin J (2015) FaceNet: a unified embedding for face recognition and clustering. In: 2015 IEEE conference on computer vision and pattern recognition (CVPR), pp 815–823. https://doi.org/10.1109/CVPR.2015.7298682

Scofano L et al (2024) Following the human thread in social navigation. arXiv: arXiv:2404.11327. Accessed 22 Sep 2024

S. C. University, Exploring ethics, social robots, and artificial intelligence. https://www.scu.edu/ethics/focus-areas/technology-ethics/resources/social-robots-ai-and-ethics/. Accessed 14 Oct 2025. [Online]

Semeraro F, Griffiths A, Cangelosi A (2023) Human-robot collaboration and machine learning: a systematic review of recent research. Rob Comput Integr Manuf 79:102432. https://doi.org/10.1016/j.rcim.2022.102432

Serrano SA, Santiago E, Martinez-Carranza J, Morales E, Sucar LE (2021) Knowledge-based hierarchical POMDPs for task planning. J Intell Robot Syst 101(4):82. https://doi.org/10.1007/s10846-021-01348-8

Sharkawy A-N, Koustoumpardis PN (2022) Human–robot interaction: a review and analysis on variable admittance control, safety, and perspectives. Machines 10(7) (Art. no. 7). https://doi.org/10.3390/machines10070591

Shi L, Copot C, Vanlanduit S (2021) GazeEMD: detecting visual intention in gaze-based human-robot interaction. Robotics 10(2) (Art. no. 2). https://doi.org/10.3390/robotics10020068

Shuchang Z (2022) A survey on human action recognition. arXiv: arXiv:2301.06082. Accessed 4 Oct 2024

Söderlund M, Natorina A (2024) Service robots in a multi-party setting: an examination of robots' ability to detect human-to-human conflict and its effects on robot evaluations. Technol Soc 77:102560. https://doi.org/10.1016/j.techsoc.2024.102560

Soori M, Arezoo B, Dastres R (2023) Artificial intelligence, machine learning and deep learning in advanced robotics, a review. Cogn Rob 3:54–70. https://doi.org/10.1016/j.cogr.2023.04.001

Spezialetti M, Placidi G, Rossi S (2020) Emotion recognition for human-robot interaction: recent advances and future perspectives. Front Rob AI 7. https://doi.org/10.3389/frobt.2020.532279

'State Machines vs Behavior Trees (2024) Designing a decision-making architecture for robotics. Accessed 5 Oct 2024. https://www.polymathrobotics.com/blog/state-machines-vs-behavior-trees

Stock-Homburg R (2022) Survey of emotions in human-robot interactions: perspectives from robotic psychology on 20 years of research. Int J Soc Rob 14(2):389–411. https://doi.org/10.1007/s12369-021-00778-6

Su H, Qi W, Chen J, Yang C, Sandoval J, Laribi MA (2023) Recent advancements in multimodal human-robot interaction. Front Neurorobot 17. https://doi.org/10.3389/fnbot.2023.1084000

Sun B et al (2016) Combining feature-level and decision-level fusion in a hierarchical classifier for emotion recognition in the wild. J Multimodal User Interf 10(2):125–137. https://doi.org/10.1007/s12193-015-0203-6

Taigman Y, Yang M, Ranzato M, Wolf L (2014) DeepFace: closing the gap to human-level performance in face verification. In: 2014 IEEE conference on computer vision and pattern recognition. IEEE, Columbus, OH, pp 1701–1708. https://doi.org/10.1109/CVPR.2014.220

Tang C, Abbatematteo B, Hu J, Chandra R, Martín-Martín R, Stone P (2024) Deep reinforcement learning for robotics: a survey of real-world successes. arXiv: arXiv:2408.03539. https://doi.org/10.48550/arXiv.2408.03539

Tatler BW, Wade NJ, Kwan H, Findlay JM, Velichkovsky BM (2010) Yarbus, eye movements, and vision. Iperception 1(1):7–27. https://doi.org/10.1068/i0382

Tomasello M, Hare B, Lehmann H, Call J (2007) Reliance on head versus eyes in the gaze following of great apes and human infants: the cooperative eye hypothesis. J Human Evol 52(3):314–320. https://doi.org/10.1016/j.jhevol.2006.10.001

Toshev A, Szegedy C (2014) DeepPose: human pose estimation via deep neural networks. In: 2014 IEEE conference on computer vision and pattern recognition, pp 1653–1660. https://doi.org/10.1109/CVPR.2014.214

Traeger ML, Strohkorb Sebo S, Jung M, Scassellati B, Christakis NA (2020) Vulnerable robots positively shape human conversational dynamics in a human-robot team. Proc Natl Acad Sci 117(12):6370–6375. https://doi.org/10.1073/pnas.1910402117

Trends in Control and Decision-Making for Human–Robot Collaboration Systems—SpringerLink. https://link.springer.com/book/10.1007/978-3-319-40533-9. Accessed 5 Oct 2024

Trombetta D, Rotithor GS, Salehi I, Dani AP (2020) Human intention estimation using fusion of pupil and hand motion. IFAC-PapersOnLine 53(2):9535–9540. https://doi.org/10.1016/j.ifacol.2020.12.2431

Truong X-T, Yoong VN, Ngo T-D (2017) Socially aware robot navigation system in human interactive environments. Intel Serv Rob 10(4):287–295. https://doi.org/10.1007/s11370-017-0232-y

Tsiourti C, Weiss A, Wac K, Vincze M (2019) Multimodal integration of emotional signals from voice, body, and context: effects of (in)congruence on emotion recognition and attitudes towards robots. Int J Soc Rob 11(4):555–573. https://doi.org/10.1007/s12369-019-00524-z

Understanding the Markov Decision Process (MDP)—built in. Accessed 5 Oct 2024. https://builtin.com/machine-learning/markov-decision-process

U. of York, AAIP launches new guidance to assure the safety of machine learned components, University of York. https://www.york.ac.uk/assuring-autonomy/news/news/amlas-published/. Accessed 14 Oct 2025. [Online]

U. of York, Assurance of machine learning for use in autonomous systems (AMLAS), University of York. https://www.york.ac.uk/assuring-autonomy/guidance/amlas/. Accessed 14 Oct 2025. [Online]

U. of York, The Safety Assurance of autonomous systems in Complex Environments (SACE), University of York. https://www.york.ac.uk/assuring-autonomy/guidance/sace/. Accessed 14 Oct 2025. [Online]

Urakami J, Seaborn K (2023) Nonverbal cues in human–robot interaction: a communication studies perspective. J Hum Robot Interact 12(2):22:1–22:21. https://doi.org/10.1145/3570169

Valori M et al (2023) Editorial: Safety in close human-robot interaction. Front Rob AI 10. https://doi.org/10.3389/frobt.2023.1288713

van de Laar T, Koudahl M, van Erp B, de Vries B (2022) Active inference and epistemic value in graphical models. Front Rob AI 9. https://doi.org/10.3389/frobt.2022.794464

Vasquez D, Fraichard T, Aycard O, Laugier C (2008) Intentional motion on-line learning and prediction. Mach Vis Appl 19(5):411–425. https://doi.org/10.1007/s00138-007-0116-9

Visca M, Kuutti SJ, Powell R, Gao Y, Fallah MS (2021) Deep learning traversability estimator for mobile robots in unstructured environments. In: Towards autonomous robotic systems 22nd annual conference, TAROS. Lincoln, UK, September 8–10, 2021. Proceedings, Springer Verlag. https://doi.org/10.1007/978-3-030-89177-0_22

Wang Z et al (2013) Probabilistic movement modeling for intention inference in human-robot interaction. Int J Rob Res 32(7):841–858. https://doi.org/10.1177/0278364913478447

Wang X, Qi X, Wang P, Yang J (2021) Decision making framework for autonomous vehicles driving behavior in complex scenarios via hierarchical state machine. Autonom Intell Syst 1(1):10. https://doi.org/10.1007/s43684-021-00015-x

Wang Y, Wang G, Ge W, Duan J, Chen Z, Wen L (2024) Perceived safety assessment of interactive motions in human–soft robot interaction. Biomimetics 9(1) (Art. no. 1). https://doi.org/10.3390/biomimetics9010058

Wiese E, Metta G, Wykowska A (2019) Robots as intentional agents: using neuroscientific methods to make robots appear more social. OSF. https://doi.org/10.31234/osf.io/5y3jz

Wu C-H, Huang Y-M, Hwang J-P (2016) Review of affective computing in education/learning: trends and challenges. Br J Educ Technol 47(6):1304–1323. https://doi.org/10.1111/bjet.12324

Wullenkord R, Eyssel F (2020) Societal and ethical issues in HRI. Curr Robot Rep 1(3):85–96. https://doi.org/10.1007/s43154-020-00010-9

Xiang X, Foo S, Zang H (2021) Recent advances in deep reinforcement learning applications for solving partially observable Markov decision processes (POMDP) problems part 2—applications in transportation, industries, communications and networking and more topics. Mach Learn Knowl Extrac 3(4) (Art. no. 4). https://doi.org/10.3390/make3040043

Xiao C, Xu L, Sui Y, Zhou R (2021) Do people regard robots as human-like social partners? Evidence from perspective-taking in spatial descriptions. Front Psychol 11. https://doi.org/10.3389/fpsyg.2020.578244

Xu X, Lu Y, Vogel-Heuser B, Wang L (2021) Industry 4.0 and industry 5.0–inception, conception and perception. J Manuf Syst 61:530–535. https://doi.org/10.1016/j.jmsy.2021.10.006

Zacharaki A, Kostavelis I, Dokas I (2021) Decision making with STPA through Markov decision process, a theoretic framework for safe human-robot collaboration. Appl Sci 11(11) (Art. no. 11). https://doi.org/10.3390/app11115212

Zaraki A et al (2017) Design and evaluation of a unique social perception system for human-robot interaction. IEEE Trans Cogn Dev Syst 9(4):341–355. https://doi.org/10.1109/TCDS.2016.2598423

Żarkowski M (2019) Multi-party turn-taking in repeated human-robot interactions: an interdisciplinary evaluation. Int J Soc Rob 11(5):693–707. https://doi.org/10.1007/s12369-019-00603-1

Zhang X, Zheng K, Wang C, Chen J, Qi H (2023) A novel deep reinforcement learning for POMDP-based autonomous ship collision decision-making. Neural Comput Appl. https://doi.org/10.1007/s00521-023-08908-z

Zhang Z, Peng G, Wang W, Chen Y, Jia Y, Liu S (2022) Prediction-based human-robot collaboration in assembly tasks using a learning from demonstration model. Sensors 22(11) (Art. no. 11). https://doi.org/10.3390/s22114279

Zhang K, Yang Z, Başar T (2021) Multi-agent reinforcement learning: a selective overview of theories and algorithms. In:Vamvoudakis KG, Wan Y, Lewis FL, Cansever D (eds) Handbook of reinforcement learning and control. Springer International Publishing, Cham, pp 321–384.https://doi.org/10.1007/978-3-030-60990-0_12

Zhao M (2023) The emotion recognition in psychology of human-robot interaction. Psychomachina 1:1–11. https://doi.org/10.59388/pm00331

Zheng M, Meng MQ-H (2012) Designing gestures with semantic meanings for humanoid robot. In: 2012 IEEE international conference on robotics and biomimetics (ROBIO), pp 287–292. https://doi.org/10.1109/ROBIO.2012.6490981

Zheng W, Wu B, Lin H (2018) POMDP model learning for human robot collaboration. In: 2018 IEEE conference on decision and control (CDC), pp 1156–1161. https://doi.org/10.1109/CDC.2018.8618904

Ziebart BD, Maas A, Bagnell JA, Dey AK (2008) Maximum entropy inverse reinforcement learning. In: Proceedings of the 23rd national conference on artificial intelligence, vol 3, in AAAI'08. AAAI Press, Chicago, IL, pp 1433–1438

Chapter 2
A Short Bibliometric Review of Metaheuristics

Mehmet Anıl Akbay, Jaume Reixach, Guillem Rodríguez Corominas, Camilo Chacón Sartori, and Christian Blum

Abstract This paper presents a concise bibliometric review of metaheuristics, a class of high-level, problem-independent algorithmic frameworks designed for heuristic optimization. Since their rise in the late twentieth century, driven largely by advances in computational power, metaheuristics have become increasingly popular in solving complex optimization problems. In this review, we examine the evolution of the metaheuristics literature, highlighting key trends, developments, and the growing impact of this field over time.

Keywords Metaheuristics · Bibliometric study · Optimization

2.1 Introduction

Given a set of available items, optimization is the process of selecting a best element, concerning some criteria. More particularly, an optimization problem is defined by a set of items, a set of constraints determining the feasible ones, an objective function, which associates a real value to each item, and the goal, which is to either maximize or minimize the objective function. In this context, a best item is called an optimal solution.

In the early days of optimization, much attention was given to developing methods to find optimal solutions to particular optimization problems. However, these methods tend to become unusable when the size of the optimization problem (the amount of available items) is increased.

The aim generally switches to finding high-quality solutions when this is the case. A solution method for a particular optimization problem that aims to find a high-quality solution without any guarantee of its optimality is called a heuristic.

M. A. Akbay · J. Reixach · G. Rodríguez Corominas · C. Chacón Sartori · C. Blum (✉)
Artificial Intelligence Research Institute (IIIA-CSIC), Bellaterra, Barcelona, Spain
e-mail: christian.blum@iiia.csic.es

B. Shukla et al. (eds.), *Sustainable Technologies for Transforming Businesses and Societies*, Transactions on Computer Systems and Networks,
https://doi.org/10.1007/978-981-96-9471-6_2

In the 1980s, the concept of metaheuristic, firstly mentioned in a seminal paper by Glover (1986), emerged as a powerful approach to optimization. Unlike traditional heuristics, which are tailored to specific problems, metaheuristics are flexible frameworks that allow the construction of heuristics for a wide range of problems. They contain a problem-dependent part, which links the metaheuristic to the problem at hand but are otherwise problem-independent. This flexibility allows metaheuristics to provide solutions across various problems without requiring extensive problem-specific adaptations.

Given the extensive and rapidly growing body of research in metaheuristics, it is important to assess and analyze the field's current state periodically. Bibliometric reviews serve as valuable tools for providing a snapshot of the current situation by analyzing publication trends and identifying key research areas, influential contributors, and shifts in geographic distribution. Such analyses can help researchers better understand the field's evolution, highlight potential opportunities for innovation, and inform future research directions.

In this paper, we present a bibliometric review of metaheuristics. We analyze a broad range of publications to provide detailed insights into the field's trends, key research areas, and influential contributions.

The remainder of the paper is structured as follows: Sect. 2.2 offers a concise overview of the key developments in the field of metaheuristics. Section 2.3 details the bibliometric analysis conducted, presenting a comprehensive examination of the literature. Finally, Sect. 2.4 summarizes the main findings and discusses their implications.

2.2 Chronology of the Main Developments in Metaheuristics

Figure 2.1 presents a chronology focused on the introduction of the main metaheuristics considered in this work and other relevant events in the field.

Metaheuristics are part of the broad field of Operations Research (OR), which focuses on finding optimal or near-optimal solutions to decision-making problems. OR emerged around World War II as a scientific approach to support decision-making through quantitative analysis. As mentioned in the introduction, the first breakthroughs in optimization consisted of the development of methods for obtaining exact solutions, such as the development of the simplex algorithm for linear programming in 1947 and of dynamic programming in 1950. Although exact methods were revolutionary, they are often limited to problems of manageable size and structure. This led to the development of the first heuristic methods, with metaheuristics emerging in the late twentieth century as more general methods that can be adapted to multiple problems. For a deeper look into the history of metaheuristics, we refer the reader to Martí et al. (2024), Sorensen et al. (2017) and Blum and Roli (2003).

Parallel to the developments depicted in Fig. 2.1, many so-called novel metaheuristics have been proposed throughout the last decades, the descriptions of which are heavily metaphor-based. These approaches often lack scientific rigor or meaningful innovation; they are based on diverse and often bizarre phenomena and offer little more than superficial analogies. They often fail to capture the metaphors they are built upon and tend to lack originality, simply rebranding existing ideas with new terminology (Camacho Villalón et al. 2020; Aranha et al. 2022). Furthermore, the experimental validation presented in the papers on these metaphor-based algorithms are often biased or poorly conducted, giving an inflated impression of their effectiveness. This trend damages the credibility of the metaheuristics research community, prompting calls for stricter editorial standards to prevent the publication of papers whose primary contribution is an ungrounded metaphor rather than a genuine scientific advance. In our bibliometric study, we ignore such approaches, even though they must be mentioned when discussing metaheuristics.

2.3 Bibliometric Study

This section presents the bibliometric study performed. The data used was obtained on `1 October 2024` from `Scopus` (www.scopus.com). To obtain such data, we decided to limit the search to a list of established metaheuristics, avoiding using more general terms that might lead to papers about the already mentioned metaphor-based metaheuristics.

The following subsection presents the methods selected and the `Scopus` search terms employed. Note that not all existing proper metaheuristics are considered, as many similarly named metaheuristic variants of great quality make an exhaustive search impossible. Every search was done using the paper's title, abstract, and keywords. In `Scopus`, this is done by using "`TITLE-ABS-KEY`."

2.3.1 Metaheuristics and Search Terms Considered

In this section, we present the metaheuristics considered in the bibliometric analysis, organized into two main categories: trajectory-based and population-based algorithms.

The trajectory-based algorithms considered in this analysis include prominent methods such as Simulated Annealing (SA), Extremal Optimization, Tabu Search (TS), Greedy Randomized Adaptive Search Procedure (GRASP), Variable Neighborhood Search (VNS), Guided Local Search (GLS), Iterated Local Search (ILS), Large Neighborhood Search (LNS), and Iterated Greedy.

On the other hand, Population-based algorithms are divided into two subcategories:

1975	Publication of the book "Adaptation in Natural and Artificial Systems" (Holland, 1975), which started the field of Evolutionary Computation.
1981	Publication of one of the first Iterated Local Search (ILS) papers (Baxter, 1981).
1983	Introduction of Simulated Annealing (Kirkpatrick et al, 1983).
1986	Introduction of Tabu Search and first use of the term "Metaheuristic" in (Glover, 1986).
1991	The first Ant Colony Optimization (ACO) variant Ant System (AS) is introduced (Colorni et al, 1991).
1995	Establishment of the Metaheuristics International Conference (MIC) and publication of the first issue of the Journal of Heuristics.
1995	Introduction of Particle Swarm Optimization (PSO) (Kennedy and Eberhart, 1995).
1995	Introduction of Greedy Randomized Adaptive Procedure (GRASP) (Feo and Resende, 1995).
1997	Introduction of Variable Neighborhood Search (VNS) (Mladenović and Hansen, 1997).
1997	Introduction of Guided Local Search (GLS) (Voudouris, 1997).
1998	Introduction of Large Neighborhood Search (LNS) (Shaw, 1998).
2000	Introduction of the physics inspired Extremal Optimization (Boettcher and Percus, 2000).
2005	Introduction of the Artificial Bee Colony (ABC) Algorithm (Karaboga et al, 2005).

Fig. 2.1 Metaheuristics chronology

- The first subcategory, Evolutionary Algorithms (EAs), includes approaches like Evolutionary Programming (EP), Genetic Algorithms (GAs), Evolution Strategies (ESs), Genetic Programming (GP), Differential Evolution (DE), and Memetic Algorithms (MAs).
- The second subcategory, Swarm Intelligence, covers methods such as Ant Colony Optimization (ACO), Particle Swarm Optimization (PSO), and Artificial Bee Colony (ABC) algorithms.

The corresponding search terms used for each category are detailed in Table 2.1.

Table 2.1 Metaheuristics and corresponding search terms considered

Category	Search terms
Trajectory-based	`TITLE-ABS-KEY(("*simulated annealing*" OR "annealing algorithm*" OR "extremal optimization" OR "tabu search" OR "taboo search" OR "greedy randomized adapt* search procedure" OR "variable neighbo* search" OR "guided local search" OR "iterated local search" OR "large neighbo* search" OR "iterated greedy"))`
Population-based evolutionary algorithms	`TITLE-ABS-KEY(("*evolutionary programming" OR "genetic algorithm*" OR "*evolution strateg*" OR "genetic programming" OR "*evolutionary comput*" OR "differential evolution" OR "memetic algorithm*"))`
Population-based: swarm intelligence	`TITLE-ABS-KEY(("ant colony optimization" OR "ant algorithm" OR "ant colony system" OR "ant system" OR "particle swarm optimization" OR "artificial bee colony"))`

2.3.2 Overview

Figure 2.2 provides an overview of the general statistics concerning the distribution of metaheuristic literature, including the number of publications by scientific area (Fig. 2.2a), publication type (Fig. 2.2b), country of origin (Fig. 2.2c), and journals (Fig. 2.2d).

In terms of scientific areas, most metaheuristic research is concentrated in Engineering and Computer Science, with over 273,000 publications. These two areas dominate the literature, reflecting their close relationship with the development and application of metaheuristic algorithms. Mathematics ranks third, with approximately half the number of publications compared to the leading fields. Following Mathematics, other areas such as Physics and Astronomy, Energy, and Materials Science have shown significant contributions. These fields are likely more application-oriented, where metaheuristic techniques are used to solve problems in the field.

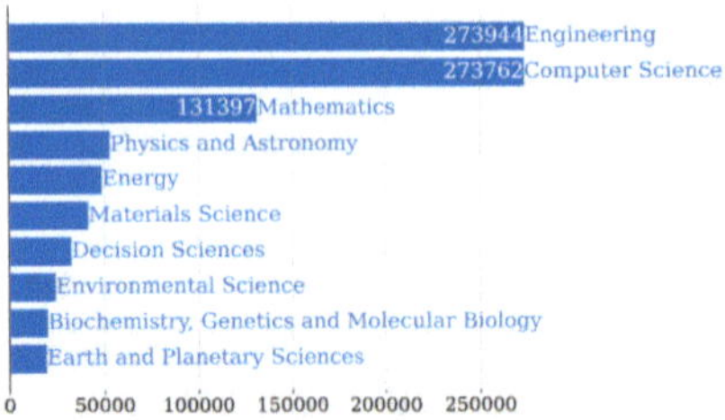

(a) Publications per scientific area

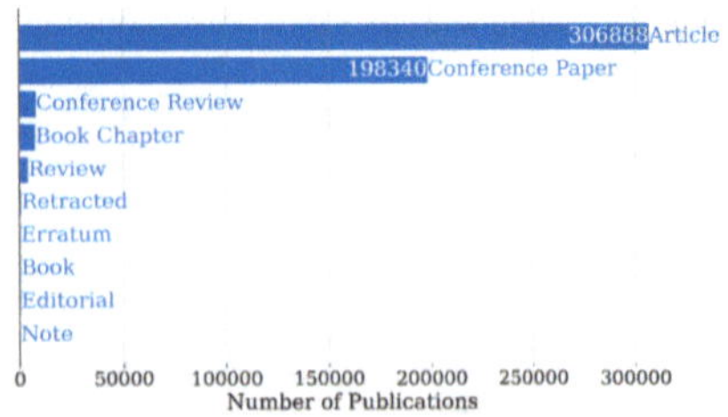

(b) Publications per paper type

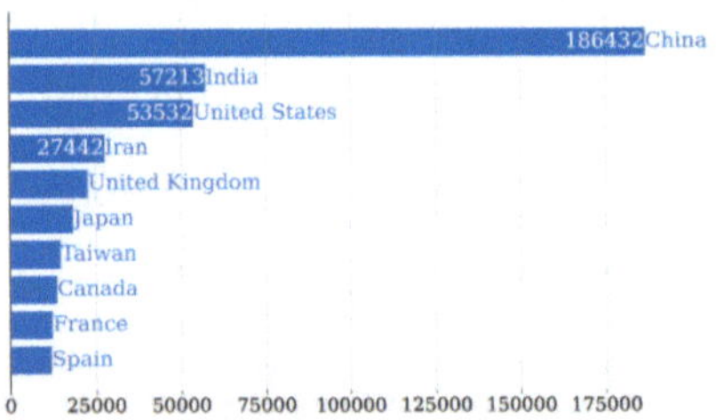

(c) Publications per country

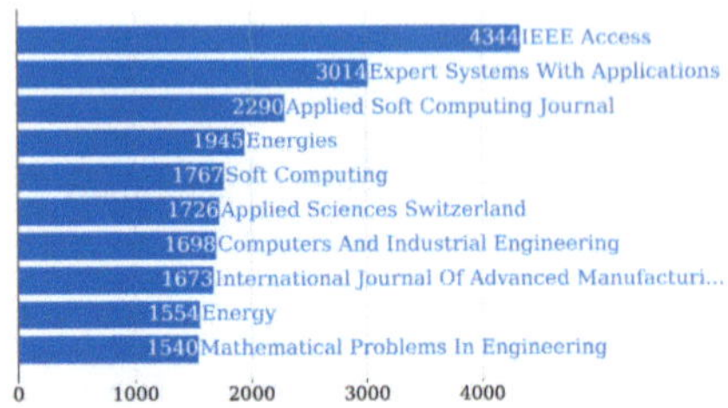

(d) Publications per journal

Fig. 2.2 Bar charts showing the distribution of metaheuristics publications across different categories: **a** publications per scientific area, **b** publications per document type, **c** publications per country, and **d** publications per journal. Only the top ten entries are shown for each category

Regarding publication types, the vast majority of research outputs are journal articles, with over 300,000 publications, indicating the preference for peer-reviewed publications. This is followed by conference papers, which account for around 200,000 publications, suggesting that conferences also play a vital role in disseminating metaheuristic research. Other publication types, such as conference reviews, book chapters, and reviews, have much smaller contributions.

Regarding national contributions, China leads by a significant margin, with over 186,000 publications in the metaheuristics field. This far exceeds the output of the second- and third-ranked countries, India and the United States, each contributing over 50,000 publications, less than one-third of China's output. Iran ranks fourth, with over 27,000 publications, about half of that of India and the United States. Other notable contributors include the United Kingdom, Japan, Taiwan, Canada, France, and Spain, with the differences among these countries being less pronounced.

When examining the leading journals in the field, IEEE Access (IEEE) ranks first, with 4344 publications. As a multidisciplinary, Q1 open-access journal, IEEE Access attracts a wide array of researchers despite being a relatively new journal established in 2013. The second most popular journal is Expert Systems with Applications (Elsevier B.V.), which has over 3000 publications. Established in 1990, this Q1 journal focuses on the application-driven design, development, and management of expert and intelligent systems across various fields, including finance, engineering, telecommunications, and medicine, to name a few. The third most prolific journal is

Applied Soft Computing (Elsevier B.V.), with 2290 publications. This Q1 journal, started in 2001, publishes research related to the application and advancement in Evolutionary Computing, Swarm Intelligence, and Neurocomputing areas, to name a few, making it highly relevant to our area of study.

2.3.3 Top Authors

Figure 2.3 presents the top 10 authors by the number of publications in the metaheuristics field overall (Fig. 2.3a), as well as in each of the three main subcategories: Evolutionary Algorithms (Fig. 2.3b), Swarm Intelligence algorithms (Fig. 2.3c), and Trajectory-based algorithms (Fig. 2.3d). However, the numbers provided should be taken with caution because it is well known that some researchers create paper mills employing large groups and collaboration networks; please note that we only refer to numbers here instead of the research quality. As in many other aspects of life, sometimes less is more.

The data highlights that M. Zhang stands out as the most prolific author in the metaheuristics field, with a total of 816 publications. Zhang's contributions are particularly concentrated in the Evolutionary Algorithms branch, where he leads with

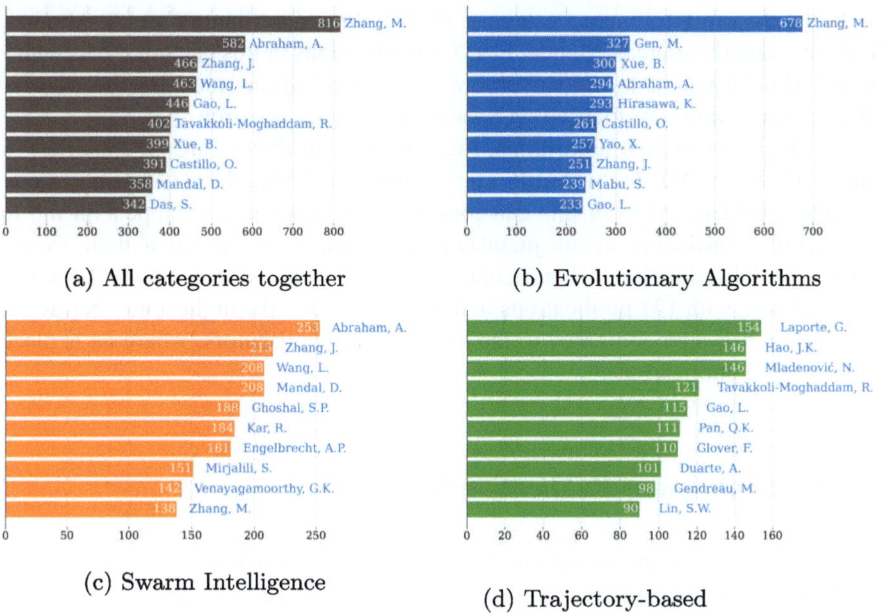

Fig. 2.3 Bar charts showing the top 10 authors in metaheuristic publications across different categories: **a** total publications, **b** publications in Evolutionary Algorithms, **c** publications in Swarm Intelligence algorithms, and **d** publications in Trajectory-based algorithms

678 publications. This represents more than 80% of his total output in the metaheuristics field, indicating a strong specialization in Evolutionary Computation. For comparison, Zhang has authored 138 publications in the Swarm Intelligence domain (10th place), which is significantly lower. Moreover, his publication count in this area is more than double that of the second-highest contributor in the category.

In second place overall is A. Abraham, with a total of 582 publications in the metaheuristics field. Abraham also leads the Swarm Intelligence category with 253 publications. However, he has more publications in the Evolutionary Algorithms category, ranking fourth with over 290 publications.

The rest of the top five general contributors are J. Zhang, L. Wang, and L. Gao. Each of these authors has over 400 publications in the field. J. Zhang and L. Wang rank second and fourth in the Swarm Intelligence category, with over 200 publications, which represents almost half of their total contributions. Additionally, L. Zhang appears in eighth place in the category of Evolutionary Algorithms. L. Gao is the tenth most prolific author in the Evolutionary Algorithms category (233 publications) and fifth in the Trajectory-based algorithms category (115 publications).

Within the Evolutionary Algorithms category, the top three contributors are M. Zhang, M. Gen (327 publications), and B. Xue (300 publications). Gen's work is highly concentrated in this category, as he does not rank within the top 10 overall contributors in the metaheuristics field. Similarly, Xue ranks seventh in the overall metaheuristics category, with approximately 75% of his publications in the Evolutionary Algorithms domain.

In the Swarm Intelligence category, the top three contributors are A. Abraham, J. Zhang, and D. Mandal. Mandal has 208 publications in this category, tied with L. Wang. Mandal also ranks ninth overall with 258 publications, meaning that over half of his contributions are focused on Swarm Intelligence research.

For Trajectory-based algorithms, the leading contributors are G. Laporte (154 publications), N. Mladenovic (146 publications), and J.K. Hao (146 publications). Despite their leadership in this category, none of these authors appear in the top 10 overall contributors in the metaheuristics field, suggesting that their work is highly specialized within this particular subfield. They are followed by R. Tavakkoli-Moghaddam, with 121 publications in the field, who has the highest number of publications in this area compared to the other top 10 contributors, where he is placed 6th with 302 publications.

2.3.4 *Evolution of Publication Statistics*

Figure 2.4 illustrates the evolution of various publication statistics over time. Specifically, Fig. 2.4a presents the overall growth of metaheuristic publications using a stacked plot, while Fig. 2.4b details this growth by algorithm category. Additionally, Fig. 2.4c and d depicts the evolution of publications in journals and conference papers, respectively. These figures use bar plots to show the percentage of each publication type (grouped in three-year intervals) relative to the total number of publications,

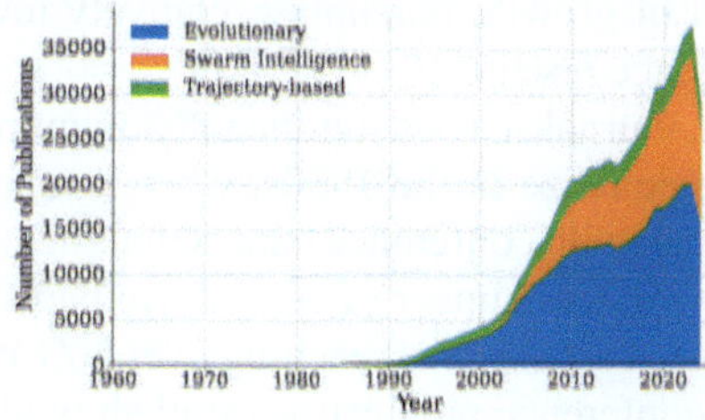

(a) All papers

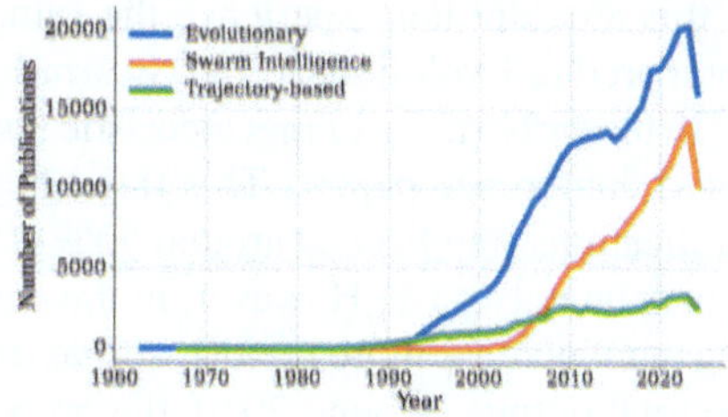

(b) All papers

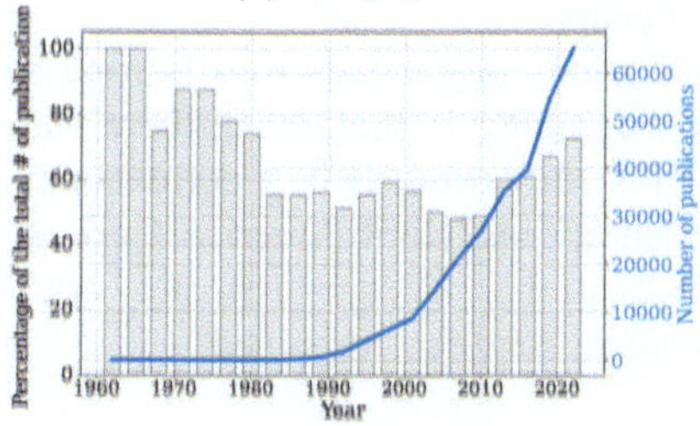

(c) Journal articles

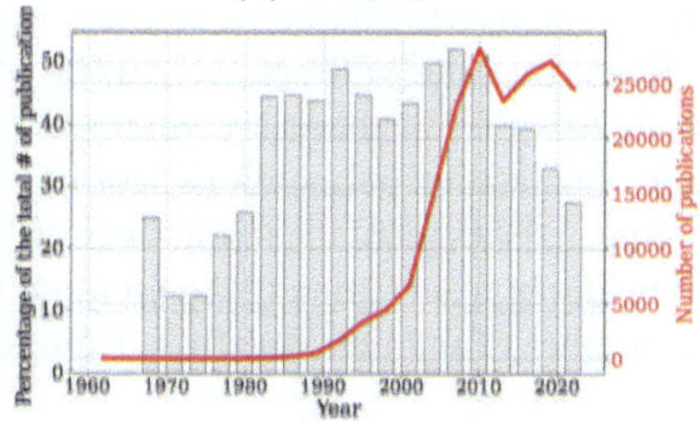

(d) Conference papers

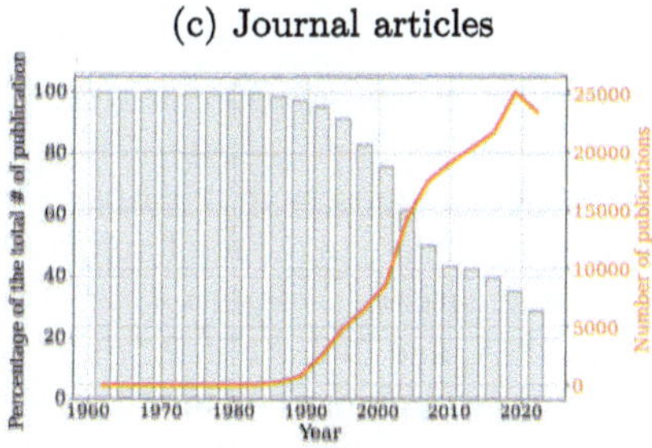

(e) Papers from EU and NA

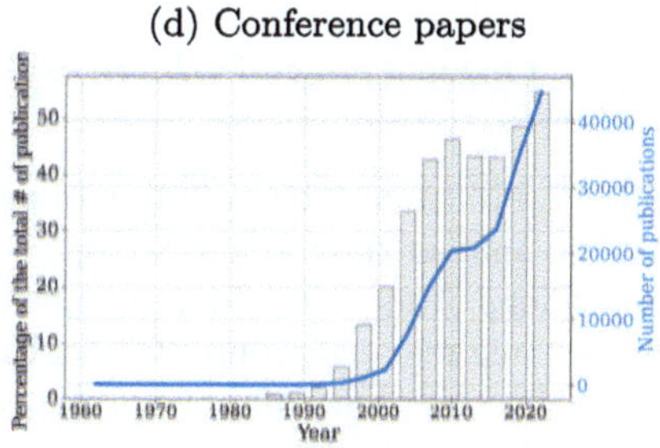

(f) Papers from China

Fig. 2.4 Plots showing the evolution of metaheuristics publications by category: **a** all papers combined, **b** papers by algorithm type (Trajectory-based, Evolutionary, and Swarm Intelligence), **c** journal articles, **d** conference papers, **e** papers from authors in Europe and North America, and **f** papers from authors in China

while lines indicate the absolute number of publications. Figure 2.4e and f focuses on the percentage of publications originating from Europe and North America (Fig. 2.4e), and China (Fig. 2.4f), following the same format as the preceding plots.

First, we can observe that the field of metaheuristics saw relatively few contributions, with the number of publications remaining low until the 1990s. During this period, research in Evolutionary Computation began to grow significantly, marking the first major increase in output. Between the 2000s and 2010s, the field experienced exponential growth, primarily driven by the rise of both Evolutionary Computation and Swarm Intelligence, the latter of which had limited output prior to this time. Despite the long history of Trajectory-based algorithms, the number of publications in this area did not experience the same level of growth, remaining relatively low compared to Evolutionary and Swarm Intelligence research.

In the early stages of metaheuristic research, journal articles significantly outnumbered conference papers. This trend began to equalize in the 1980s, when journal publications stabilized at around 50% of the total, with conference papers making up approximately 45%. However, in the early 2000s, the number of conference papers began to increase more rapidly, surpassing journal articles and reaching over 50% of the total output. Around 2010, the growth of conference publications stalled, while journal papers continued to increase, leading to a rise in the proportion of journal articles relative to conference papers in recent years.

Regarding the authors' origin, most publications originated from Europe and North America in the early years. However, beginning in the 2000s, there was a rapid increase in publications from China. This shift resulted in a corresponding decline in the percentage of publications from Europe and North America. Currently, publications from China make up more than 50% of the total, approximately double the number of publications from Europe and North America combined.

2.3.5 *Open-Access Publishing*

Figure 2.5 illustrates the evolution of open-access publications, both in terms of the absolute number and the share of open-access papers relative to the total number of publications. This is presented for the general metaheuristics literature (Fig. 2.5a) and for three regional groups: China, Europe and USA, and India (Fig. 2.5b).

Initially, the proportion of open-access publications remained relatively low, at around 10% of the total output, until approximately 2010. Since then, there has been a rapid increase in the adoption of open access, with the share rising to over 30% in recent years. However, the use of open access varies significantly among different regions. Europe and USA consistently show a higher share of open-access publications, exceeding 25% over time, while the global average is around 20%. In contrast, China has a lower proportion of open-access publications, with figures below 20%, and India exhibits a slightly lower share, just above 15%, both of which fall below the global mean.

2.3.6 *Keywords Analysis*

Figure 2.6 presents the most frequently used keywords across each metaheuristic category. Figure 2.6a displays the top 5 keywords along with the number of publications that make use of them, visualized as a bar plot. Figure 2.6b uses a Venn diagram to illustrate the keywords, highlighting how these keywords are shared across different categories. For clarity, each keyword is assigned an ID in alphabetical order, and the relationship between these IDs and the corresponding keywords is provided in the figure caption. Note that some keywords are repeated despite their similarity, as merging similar terms consistently pose a challenge, and we present them as they appear indexed in the Scopus database.

The data reveals two main keywords consistently ranked within the top 5 for all three categories: "12: Optimization" (ranked 3rd in both Evolutionary Algorithms and Swarm Intelligence, and 2nd in Trajectory-based methods) and "1: Algorithms" (5th in Evolutionary Algorithms and Swarm Intelligence, and 4th in Trajectory-based methods). Additionally, the keyword "7: Genetic Algorithms" appears in the top 10 across all three categories. While this is expected in the Evolutionary Algorithms domain, its prominence in other categories likely comes from Genetic Algorithms being one of the most widely used metaheuristics, frequently used as a comparison benchmark in studies from other areas. It should also be noted that publications may overlap between categories, as some studies compare algorithms across different fields.

Focusing on each category individually, the most used keywords in the Evolutionary Algorithms category are Genetic Algorithms (6 and 7), reflecting the dominant position of these techniques in this field. The fourth most used keyword, "4: Evolutionary Algorithms," indicates that searching solely for this term would not retrieve all relevant papers, as not all studies make use of it. Other notable keywords include "10: Multiobjective," which likely corresponds to research in Multi-Objective Evolutionary Algorithms (MOEAs), a significant area within the field, with prominent algorithms such as MOEA/D (Multi-Objective Evolutionary Algorithm Based on Decomposition) or NSGA-II (Non-dominated Sorting Genetic Algorithm II), to name a few. Furthermore, "11: Neural Networks" suggests the presence of research related to Neuroevolution, where evolutionary algorithms are employed to optimize neural networks, either in terms of their weights or topology.

In the Swarm Intelligence category, the top keywords are associated with Particle Swarm Optimization (PSO) (15, 14, 13), as expected due to PSO's widespread use in this area. Completing the top five, we have "2: Ant Colony Optimization," although the number of papers featuring this term is significantly lower compared to PSO.

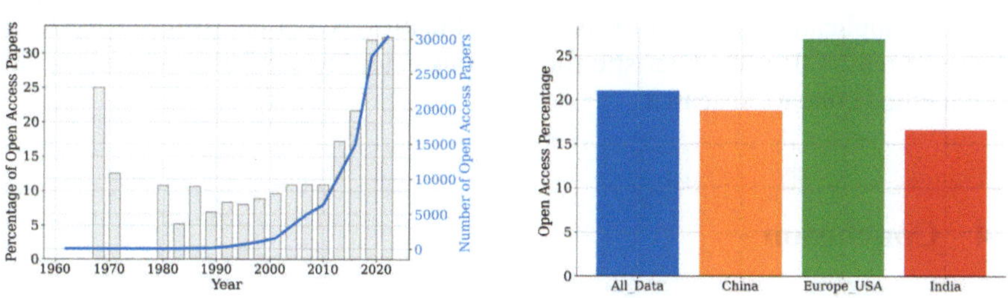

(a) Trend

(b) Open Access Percentage

Fig. 2.5 Plots showing the evolution of **a** open-access publications and **b** the share of open-access publications relative to the total number of publications in each category (all publications, publications from China, Europe, USA, and India)

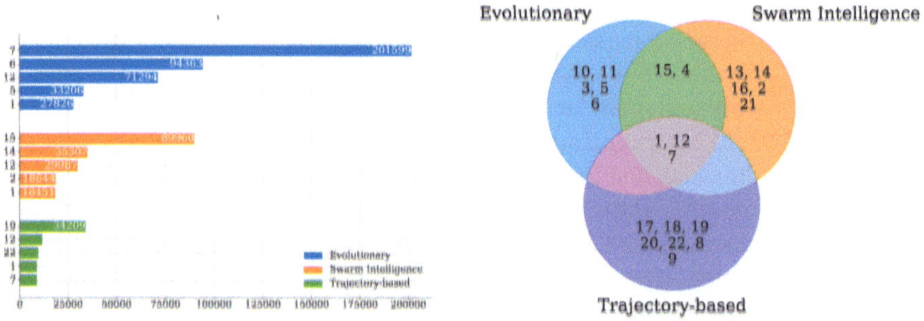

(a) Top 5 keywords per category (b) Top 10 keywords for each category

Fig. 2.6 Top keywords per categories. The numbers in each section corresponds to the following keywords: 1—algorithms, 2—ant colony optimization, 3—article, 4—artificial intelligence, 5—evolutionary algorithms, 6—genetic algorithm, 7—genetic algorithms, 8—heuristic algorithms, 9—heuristic methods, 10—multiobjective optimization, 11—neural networks, 12—optimization, 13—particle swarm, 14—particle swarm optimization, 15—particle swarm optimization (PSO), 16—particle swarm optimization algorithm, 17—problem solving, 18—scheduling, 19—simulated annealing, 20—simulated annealing algorithms, 21—swarm optimization, 22—Tabu search

Additionally, "21: Swarm Optimization" appears in the top 10, indicating that broader terms may miss a significant portion of the literature, as general category names are not commonly used for indexing.

Finally, in the Trajectory-based category, "19: Simulated Annealing" ranks as the most used keyword, nearly doubling the occurrence of the next most frequent terms. "22: Tabu Search," another well-established technique, ranks third. Interestingly, "17: Problem Solving" appears within the top 10 keywords, suggesting that research in this category is often more application-driven, focusing on how these methods can be applied to specific problems rather than on further developments in the algorithms themselves. This may be attributed to the fact that many of these techniques, such as Simulated Annealing and Tabu Search, were introduced in the 1980s and are now considered well-established. "18: Scheduling," another top keyword, further emphasizes this application-oriented focus. Lastly, the presence of heuristic-related terms (8 and 9) may reflect the heuristic roots of this category and its strong association with methods like local search.

2.4 Conclusion

This paper presents a comprehensive bibliometric review of metaheuristics, offering a detailed analysis of the field's evolution and prevailing trends. Our study categorized metaheuristics into two primary groups: Trajectory-based algorithms and Population-based algorithms, the latter further divided into Evolutionary Algorithms and Swarm Intelligence methods. By focusing on key aspects such as the chronology of developments, scientific domains, publication types, top countries, leading journals, and prolific authors, we have outlined the trajectory of metaheuristics research over the decades.

Our analysis highlights that early research in metaheuristics focused on foundational algorithms like Simulated Annealing, Tabu Search, and various Evolutionary Algorithms. These seminal contributions laid the groundwork for metaheuristics to become vital tools for addressing complex optimization problems across multiple domains. The steady growth of research in Evolutionary Computation and Swarm Intelligence reflects the increasing recognition of their effectiveness and flexibility. Although Trajectory-based metaheuristics have a longer history, their publication growth has been comparatively modest, suggesting a potential area for renewed research efforts. A significant observation from our bibliometric study is the geographical shift in metaheuristics research. Once dominated by Europe and North America, the field now sees China leading in the number of publications, with India emerging as a notable contributor. This shift might be driven by increased research funding, strategic focus on technological advancement, and academic expansion in these regions. We also noted a change in publication patterns. Conference papers played a leading role in the early stages of metaheuristics research, but journal publications have become prominent in recent years. This trend could indicate the increasing importance of journal articles within academia and funding bodies. Another key trend is the rise of open-access publications, especially in Europe and North America, although open-access adoption remains lower in China and India. Expanding global access to research findings remains a priority to ensure broader dissemination of knowledge.

An interesting trend that warrants further investigation is the proliferation of so-called novel metaphor-based metaheuristics, which lack the scientific rigor traditionally expected in this field. While not the focus of this paper, the significant increase in such publications, particularly in specific research communities, raises concerns about the integrity and future direction of metaheuristics research. A thorough analysis of their impact and distribution across the research landscape would provide valuable insights into the health and sustainability of the field.

In conclusion, our bibliometric review provides valuable insights into the development and current state of the research in metaheuristics. By understanding the trends and shifts within the field, researchers can better navigate the evolving landscape of optimization techniques and contribute to advancing this critical study area.

Acknowledgements The research presented in this paper was supported by grants TED2021-129319B-I00 and PID2022-136787NB-I00 funded by MCIN/ AEI/10.13039/501100011033.

References

Aranha C, Camacho Villalón CL, Campelo F, Dorigo M, Ruiz R, Sevaux M, Sörensen K, Stützle T (2022) Metaphor-based metaheuristics, a call for action: the elephant in the room. Swarm Intell 16(1):1–6

Baxter J (1981) Local optima avoidance in depot location. J Oper Res Soc 32(9):815–819

Blum C, Roli A (2003) Metaheuristics in combinatorial optimization: overview and conceptual comparison. ACM Comput Surv 35(3):268–308

Boettcher S, Percus A (2000) Nature's way of optimizing. Artif Intell 119(1–2):275–286

Camacho Villalón CL, Stützle T, Dorigo M (2020) Grey wolf, firefly and bat algorithms: three widespread algorithms that do not contain any novelty. In: International conference on swarm intelligence. Springer, pp 121–133

Colorni A, Dorigo M, Maniezzo V (1991) Distributed optimization by ant colonies. In: Proceedings of ECAL 1991—first European conference on artificial life, vol 142. Elsevier, Paris, pp 134–142

Feo TA, Resende MG (1995) Greedy randomized adaptive search procedures. J Global Optim 6:109–133

Glover F (1986) Future paths for integer programming and links to artificial intelligence. Comput Oper Res 13(5):533–549

Holland JH (1992) Adaptation in natural and artificial systems, 2nd edn. University of Michigan Press, Ann Arbor, MI (1975)

Karaboga D et al (2005) An idea based on honey bee swarm for numerical optimization. Technical report-tr06, Erciyes university, engineering faculty, computer ..

Kennedy J, Eberhart R (1995) Particle swarm optimization. In: Proceedings of ICNN'95-international conference on neural networks, vol 4. IEEE, pp 1942–1948

Kirkpatrick S, Gelatt CD Jr, Vecchi MP (1983) Optimization by simulated annealing. Science 220(4598):671–680

Martí R, Sevaux M, Sörensen K (2024) 50 Years of metaheuristics. Eur J Oper Res

Mladenović N, Hansen P (1997) Variable neighborhood search. Comput Oper Res 24(11):1097–1100

Shaw P (1998) Using constraint programming and local search methods to solve vehicle routing problems. In: International conference on principles and practice of constraint programming. Springer, pp 417–431

Sorensen K, Sevaux M, Glover F (2017) A history of metaheuristics. arXiv preprint arXiv:1704.00853

Voudouris C (1997) Guided local search for combinatorial optimisation problems. Ph.D. thesis, University of Essex

Chapter 3
Queuing for Civility: Regulating Emotions and Reducing Toxicity in Digital Discourse

Akriti Verma, Shama Islam, Valeh Moghaddam, and Adnan Anwar

Abstract The pervasiveness of online toxicity, including hate speech and trolling, disrupts digital interactions and online well-being. Previous research has mainly focused on post-hoc moderation, overlooking the real-time emotional dynamics of online conversations and the impact of users' emotions on others. This paper presents a graph-based framework to identify the need for emotion regulation within online conversations. This framework promotes self-reflection to manage emotional responses and encourage responsible behaviour in real time. Additionally, a comment queuing mechanism is proposed to address intentional trolls who exploit emotions to inflame conversations. This mechanism introduces a delay in publishing comments, giving users time to self-regulate before further engaging in the conversation and helping maintain emotional balance. Analysis of social media data from Twitter and Reddit demonstrates that the graph-based framework reduced toxicity by 12%, while the comment queuing mechanism decreased the spread of anger by 15%, with only 4% of comments being temporarily held on average. These findings indicate that combining real-time emotion regulation with delayed moderation can significantly improve well-being in online environments.

Keywords Digital emotion regulation (DER) · Interpersonal emotion regulation (IER) · Emotions in social media · Emotions online · Human–computer interaction (HCI) · Affective computing

---

A. Verma (✉) · S. Islam
School of Engineering, Deakin University, Geelong, VIC, Australia
e-mail: vermaakr@deakin.edu.au

S. Islam
e-mail: shama.i@deakin.edu.au

V. Moghaddam · A. Anwar (✉)
School of Information Technology, Deakin University, Geelong, VIC, Australia
e-mail: adnan.anwar@deakin.edu.au

V. Moghaddam
e-mail: valeh.moghaddam@deakin.edu.au

B. Shukla et al. (eds.), *Sustainable Technologies for Transforming Businesses and Societies*, Transactions on Computer Systems and Networks,
https://doi.org/10.1007/978-981-96-9471-6_3

3.1 Introduction

The fast pace of online interactions often leads to emotionally charged conversations, increasing the potential for polarised reactions and digital conflict (Smith et al. 2022). The rising concern of online toxicity, which often includes hate speech and trolling, has negative effects on users' well-being and the overall health of digital communities (Goel et al. 2016; Maarouf et al. 2210). Current methods for managing online toxicity primarily focus on post-hoc moderation, where harmful content is identified and removed after it's been posted (Chandrasekharan et al. 2022). Despite advancements in content moderation tools, there is still a need for proactive solutions that empower users to manage their emotional impact on conversations before toxicity escalates (Gongane et al. 2022; Trujillo and Cresci 2022).

Digital Emotion Regulation (DER), the use of digital technologies to influence one's emotional state, has recently emerged as a habit individuals acknowledge and employ in everyday life by combining a variety of applications and devices for purposefully managing emotions (Smith et al. 2022; Wadley et al. 2020). Some examples include listening to uplifting music while exercising, watching comedy or light-hearted videos to relieve stress after work, playing social video games when lonely, or scrolling through social media applications to combat boredom. While DER applications are widely acknowledged in personal contexts, they are underexplored in interactive online environments where emotions can propagate quickly and influence others (Manikonda et al. 1803). Additionally, existing solutions inadequately address the gap between reactive moderation and proactive emotional regulation. This paper introduces a new approach to promoting healthier online conversations by advocating self-reflection-based emotion regulation and integrating it with a comment queuing mechanism to curb the propagation of negative emotions, particularly in conversations vulnerable to trolling.

The practice of self-reflection, which involves assessing one's own emotions and behaviour, has been widely acknowledged as a method for managing emotional well-being in everyday interactions (Herwig et al. 2010). However, its potential for regulating emotions in online settings has not been thoroughly explored. By encouraging self-reflection in digital spaces, users can gain greater awareness of how their comments affect conversations emotionally. This will allow them to re-evaluate and manage their emotions before engaging further (Kiskola et al. 2021). This proactive approach shifts the focus from reactive moderation, which deals with toxicity only after it has occurred, to real-time emotion regulation, thereby reducing the chances of emotional escalation.

This paper introduces a method that utilises a graph-based framework to visualise the emotional tone of a conversation and informs users of their impact on its emotional state. This approach promotes self-reflection, providing users with an opportunity to consider the emotional impact of their comments. Additionally, in cases of deliberate trolling, a comment queuing mechanism is proposed. This system introduces a brief delay in publishing comments, allowing users to reflect on their emotional state during the pause, while also ensuring the emotional balance of the conversation is maintained.

Our analysis of social media data shows that promoting online self-reflection can be a powerful tool for enhancing emotion regulation, especially in navigating emotionally charged conversations. This paper highlights the value of Implicit Emotion Regulation through Self-Reflection and deliberately delayed responses by offering an opportunity for cognitive reappraisal (Kiskola et al. 2022). This subtly nudges users towards a more considerate and regulated response, aiming to reduce online toxicity and facilitate healthier digital spaces. Therefore, this work makes the following research contributions:

- A novel self-reflection-based DER system: This paper proposes a graph-based framework that informs users of their emotional influence on a conversation and promotes implicit emotion regulation through a user-centred approach.
- A comment queuing mechanism for real-time emotion regulation: This paper introduces a comment queuing system that delays user responses to prevent impulsive emotional reactions, offering users time to reflect on their emotional state. Unlike existing delay mechanisms, our system is adaptive and context-sensitive, dynamically adjusting based on the emotional dynamics of the conversation.
- Empirical validation of emotion regulation in digital conversations: Our initial findings show a 12% reduction in the spread of hate speech and anger, exceeding Google's Perspective API by 3%. Furthermore, when using the queue mechanism during ongoing conversations, there is a potential decrease of up to 15% in trolling and hate speech propagation, with only 4% of comments temporarily held for an average of 47 s.

3.2 Literature Review

As social media platforms grow, there is an increasing need to address emotional well-being in online environments. This brief literature review explores recent research on DER, self-reflection as a tool for emotion regulation, trolling behaviour, and strategies for comment moderation. It highlights the significance of this research in advancing our understanding of these areas.

3.2.1 Digital Emotion Regulation (DER)

DER is a growing field of study within psychology and human–computer interaction, focusing on the influence of digital tools on individuals' emotional well-being. DER research examines the impact of digital technologies on emotion regulation (ER) (Gross 2014, 2015; McRae and Gross 2020; Wadley et al. 2020). Studies have utilised methods such as self-reports and diaries to demonstrate the use of digital technologies for everyday ER (Hossain et al. 2022; Lukoff et al. 2018; Shen and Cox 2020; Smith et al. 2022). DER interventions have incorporated biofeedback and reminder-based systems to enhance ER skills by encouraging users to practice ER (Kiskola et al. 2021; Kou and Gui 2020). Additionally, multi-modal sensors, such as cameras and touch sensors, have been employed to observe and recognise emotional changes during the DER process (Luo and Yang 2021; Ruensuk et al. 2020). Research has also revealed the concentration of toxic discussions, particularly among individuals with limited social connections (Saveski et al. 2021; Thomas et al. 2022). Furthermore, studies have shown that moral emotions expressed in tweets can influence the circulation of false rumours, with key users playing a crucial role in initiating online social movements (Solovev and Pröllochs 2022). However, much of the existing literature focuses on individual emotional regulation and overlooks the social dynamics of online conversations, where emotional contagion from others can significantly impact one's emotional state. This study addresses the gap in current DER literature by introducing an approach that includes emotion regulation strategies and feedback mechanisms during online interactions.

3.2.2 Emotion Regulation and Self-reflection

Emotion regulation, as defined by Gross (2008), encompasses the processes through which individuals manage the emotions they experience, how they express them, and the timing of these emotions. Traditionally studied in offline environments, emotion regulation enhances interpersonal communication by enabling individuals to reconsider their emotional responses before reacting. Self-reflection is a crucial component of emotion regulation that empowers individuals to assess their emotions and underlying motivations, potentially leading to cognitive reappraisal (Ochsner and Gross 2005). Studies have found that temporal distancing—taking time before responding to emotional stimuli—can decrease emotional arousal and promote better emotional control (Grossmann and Kross 2014). While this concept has mainly been explored in face-to-face and other offline settings, it is seldom put into practice in digital interactions where the rapid pace of interaction often allows little time for reflection. Existing research has examined emotion regulation in digital contexts but primarily within educational tools or therapeutic applications, with minimal attention to public online discourse (Hadwin et al. 2017, 2019). Our work bridges this gap by developing a framework that promotes self-reflection in real-time online conversations, enabling users to pause, reflect, and regulate their emotions before posting a potentially toxic comment.

3.2.3 Online Toxicity and Deliberate Trolling

While labelling emotions may help control unintentional emotional outbursts, dealing with deliberate trolling is more complex (Jane 2020). Trolling, which involves intentionally disrupting online conversations to provoke emotional reactions, has been extensively studied in social and computer science (Buckels et al. 2014). Studies have identified trolls as individuals who derive pleasure from causing emotional chaos and escalating conflict for a range of motivations from seeking amusement and attention to more malicious intentions such as causing harm or manipulating discussions (Cheng et al. 2017; Kumar et al. 2017). Current moderation approaches primarily focus on removing harmful content after it is posted, which is reactive rather than proactive. While tools such as banning a user and deletion of comments can mitigate some of the damage caused by trolls, they do not address the initial escalation (Chandrasekharan et al. 2022, 2015; Yin et al. 2009). Prior research suggests that traditional content moderation often exacerbates trolling behaviour, as trolls enjoy pushing the boundaries of platform policies (Phillips 2015). This highlights the need for a more preventive approach to managing trolls, which our work aims to achieve through a comment queuing mechanism designed to allow for self-reflection-based emotion regulation.

3.2.4 Conversation Analysis and Comment Moderation Mechanisms

Comment moderation has been the primary defence against online toxicity, typically employing the detection and removal of harmful content, as well as flagging inappropriate behaviour. Platforms such as Facebook and Twitter have implemented algorithmic tools, such as content filters and moderation bots, to mitigate toxic behaviour (Gillespie 2020; Gorwa et al. 2020). However, recent literature has highlighted their limited capacity to effect meaningful, long-term changes in user conduct (Gillespie 2018). It has been argued that the root of the problem lies not just in identifying harmful content but in changing the emotional responses and behaviours that drive users to post it in the first place (Frischlich et al. 2021; Park et al. 2022).

In digital communication, effective emotion regulation (ER) includes the utilisation of factual cues, automatic emotion identification, and didactic learning (Kiskola et al. 2021; Slovak et al. 2204). Implicit emotion regulation, which operates automatically and can be automated, has garnered recent attention. Affect labelling, making emotionally charged aspects of conversations more apparent, can enhance emotion regulation (Torre and Lieberman 2018). Current research has explored approaches to comment moderation, such as the introduction of delays aimed at limiting impulsive reactions. For example, a delayed feedback loop can provide users time for reconsideration before posting their comments (Cheng et al. 2017). This aligns with emotion regulation theory, which argues that the introduction of time-based pauses can facilitate users in engaging in cognitive reappraisal, thereby defusing emotionally charged situations. Yet, most existing comment delay mechanisms are either arbitrarily timed or overlook the emotional dynamics in the conversation.

Therefore, the following gaps exist in the literature:

- Lack of Real-Time Emotion Regulation in Online Conversations: Current research explores Digital Emotion Regulation (DER) in personal use of digital tools, but lacks real-time strategies for regulating emotions in interactive online settings, focusing mainly on post-hoc content moderation rather than preventing emotional escalation during conversations.
- Limited Application of Self-Reflection in Digital Spaces: Although self-reflection is commonly used for offline emotion regulation, its application to online communication is not well explored. Methods to integrate self-reflection into digital interactions can promote more thoughtful, emotionally regulated responses.
- Limited Focus on the Root Causes of Trolling: Current research focuses on removing or moderating toxic content rather than addressing the root causes of trolling behaviour, such as emotional arousal and impulsivity. Strategies are needed to encourage users to pause and reflect before posting.

Our research extends upon these findings by integrating a comment queuing system with a graph-based framework designed to monitor the emotional tone of the conversation continually. This approach not only introduces temporal delays to encourage self-reflection but also ensures that the emotional equilibrium of the conversation is maintained, thereby reducing the probability of escalation resulting from trolling or impulsive behaviour.

3.3 Methodology

This paper presents an approach to managing emotions in online conversations through self-reflection. The methodology comprises four stages of the eImpact framework (Fig. 3.1), including emotion detection, graph-based conversation analysis, and the implementation of a comment queuing system to encourage self-reflection and minimise emotional escalation.

3.3.1 Data Collection and Preprocessing

For this study, we focused on data from Twitter and Reddit due to their distinct communication styles and ability to capture a wide range of emotional expressions.

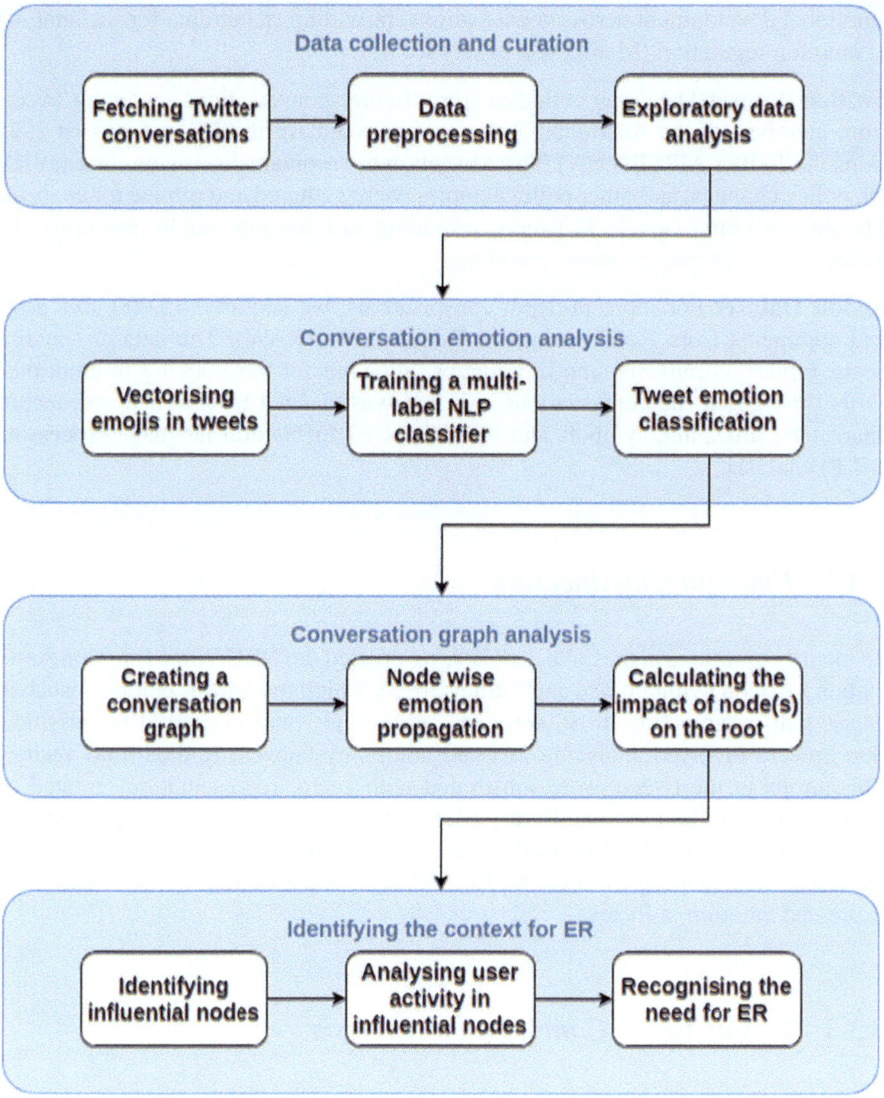

Fig. 3.1 eImpact: framework for supporting emotion regulation in social media conversations

Twitter facilitates rapid, public interactions that often lead to direct and emotionally charged conversations, particularly around political and social issues. Its fast-paced nature makes it well-suited for analysing how emotions propagate and the need for self-reflection in digital interactions (Manikonda et al. 1803).

On the other hand, Reddit, with its longer posts and topic-specific communities, allows for more in-depth discussions. Its threaded structure enables the tracking of emotional development across conversations, providing richer data for the analysis of emotion regulation (Manikonda et al. 1803).

Twitter Dataset Data was collected from Twitter conversations involving tweets from members of the Australian Parliament between April 2020 and August 2022 using the Twitter API (Tweepy). These tweets, which contain discussions on COVID-19, policy changes, and other political topics, were gathered and refined for analysis. The dataset comprises 25 K tweets, providing insights into public reactions and interactions expressing various emotions.

Reddit Dataset For more in-depth conversations, we acquired 15,000 user posts and comments from Reddit using the Reddit API (PRAW). The dataset contains posts, replies, comments, and timestamps, allowing for the tracking of emotional shifts throughout the conversations. All text was cleaned to eliminate extraneous characters, links, and symbols and was tokenized for natural language processing (NLP) analysis.

3.3.2 Emotion Classification

To identify emotions in the collected data, we utilised the NRC Word-Emotion Association Lexicon (Mohammad and Turney 2013), which recognises emotions such as anger, fear, anticipation, trust, surprise, sadness, joy, and disgust. The classification process involved analysing text and emojis by converting them into vectors. The emojis in the tweets were substituted with vector representations created by Gensim through the Emojinal library (Barry et al. 2021), after which the tweet text was tokenized using the TweetTokenizer. Each comment received an emotion and intensity score ranging from 0.1 to 1.0, reflecting the emotion and strength of the expressed emotion in its text.

3.3.3 Graph-Based Conversation Analysis

Conversations were represented using a directed acyclic graph (DAG) structure, where original posts function as root nodes, while replies and comments act as child nodes. The connections between comments and replies are represented by edges. This hierarchical structure is then used to assess the influence of comments on the overall conversation.

The eImpact model represents every comment or tweet as a node in the graph, with each node being allocated an emotion score (Verma et al. 2023). This score is determined by the likelihood that the emotion classification model associates with the content. The influence of each node on the root node (original post) is evaluated using various metrics:

- Number of Replies: Nodes with more replies are considered more influential.
- Distance from Root Node: Nodes closer to the original post have a higher influence on the overall emotional tone.
- PageRank: A comment's importance is determined by its position in the graph and the volume of engagement it receives.
- Emotion Intensity: Each comment's emotional intensity contributes to its influence score, affecting the root node's overall emotion board.
- This approach enables the model to identify specific nodes within the conversation likely to trigger emotional escalation or toxicity.

This graph-based structure allows us to assess how each comment influences the emotional tone of the conversation. These influence scores are used to update the emotion board of the root node, which compiles the emotional impact of all comments. To prevent the escalation of negative emotions, highly toxic comments are temporarily held before being added to the conversation graph, based on their impact on the root node.

3.3.4 Comment Queuing System for Self-reflection

To facilitate self-reflection and regulate negative emotions in online conversations, we introduce a comment queuing mechanism as shown in Fig. 3.2. This system assesses each new comment to calculate its potential impact on the overall emotional tone of the conversation before its publication. As each comment influences the emotional tone of the conversation, its inclusion also means an addition of emotional weight to the cumulative tone of the conversation. If the emotional impact of a comment pushes the predefined thresholds, such as Anger > 50% or Fear > 60%, the comment is flagged as toxic and temporarily stored in a queue. For instance, as shown in Fig. 3.5, in a Twitter conversation centred around government policy reforms, a reply that expressed strong anger towards a government policy raised the root node's anger threshold to 65%, Fig. 3.3. As a result, it was queued (indicated in red). This comment, positioned directly under the root node, significantly impacted the overall sentiment due to its emotional intensity. However, as can be seen in Fig. 3.4, when subsequent comments introduced feelings of trust and joy, the anger percentage fell below 55%, enabling the queued comment to be integrated safely without further escalation (Fig. 3.5).

While in the queue, comments are regularly re-evaluated each time a new comment is added to the conversation. This guarantees that the emotional board is continuously updated, enabling the system to determine whether previously queued comments can now be added without surpassing the thresholds. If the addition of a new comment balances or reduces the overall emotional intensity, the previously flagged comment can be safely integrated into the conversation.

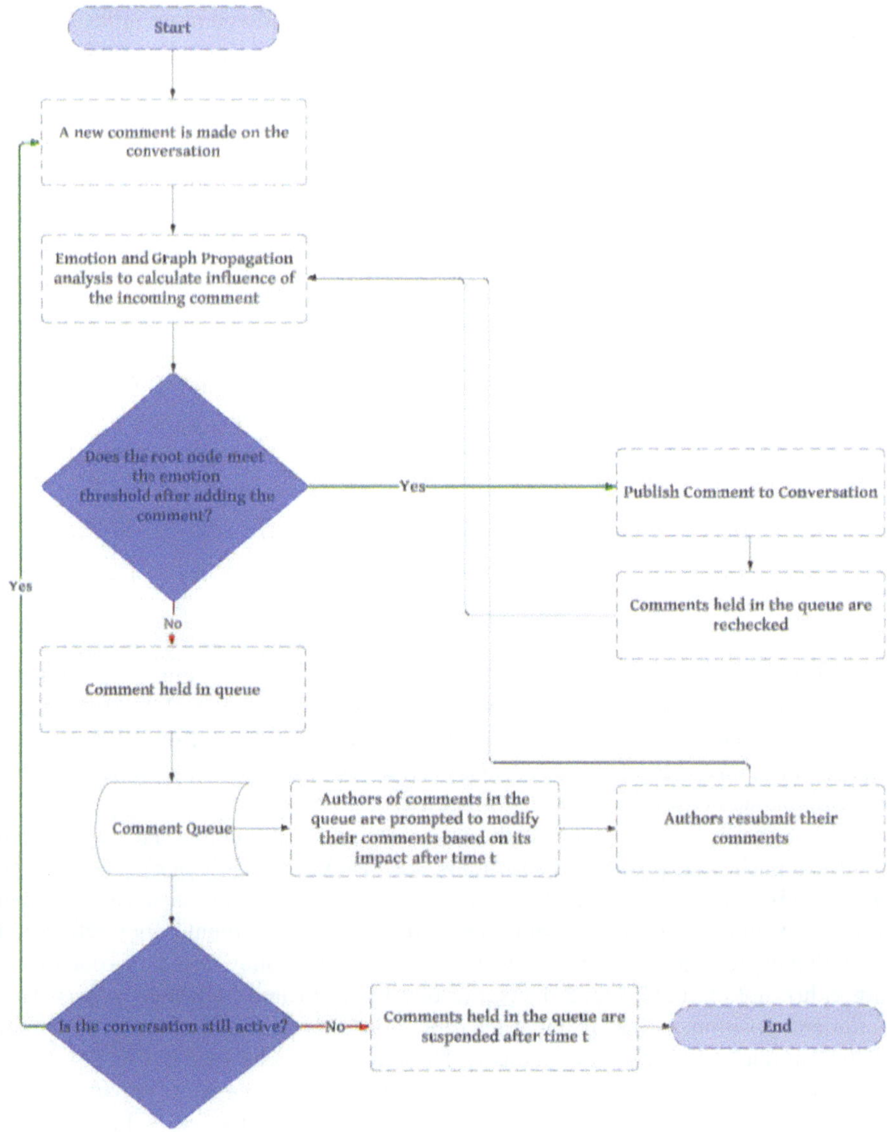

Fig. 3.2 Proposed comment queuing to encourage self-reflection

To accommodate the flow of online conversations, the system distinguishes between active and non-active conversation stages. During active stages, the thresholds are more strict to prevent the escalation of negative emotions. As the conversation becomes less active, these thresholds may be relaxed, permitting more comments to enter the conversation while still upholding its emotional balance.

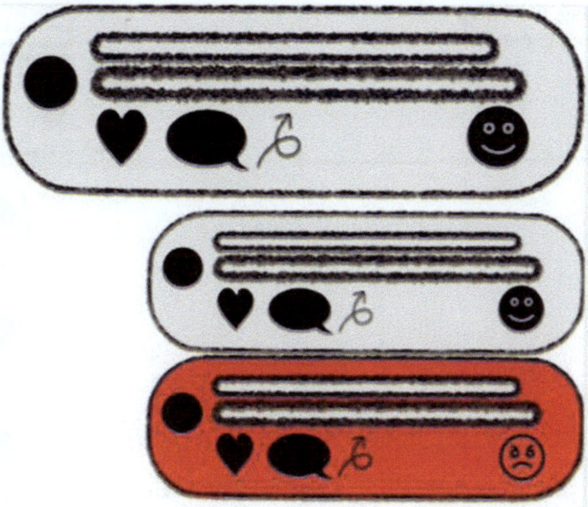

Fig. 3.3 Angry comment queued

If a comment remains in the queue after all the other comments have been processed and still exceeds the emotional thresholds, its author is prompted to revise the comment. Following revision, the comment is re-evaluated. If the revised comment aligns with the emotional thresholds, it is included in the conversation; if not, the comment is suspended to prevent further emotional escalation and ensure a healthier dialogue.

The threshold values for the emotional tone of the conversation are determined using several parameters and are adjusted dynamically through an algorithm that takes into account the volume of comments, the general emotional distribution in the conversation, and recent variations in emotional intensity. For example, in very active discussions where numerous comments express high-intensity feelings, the

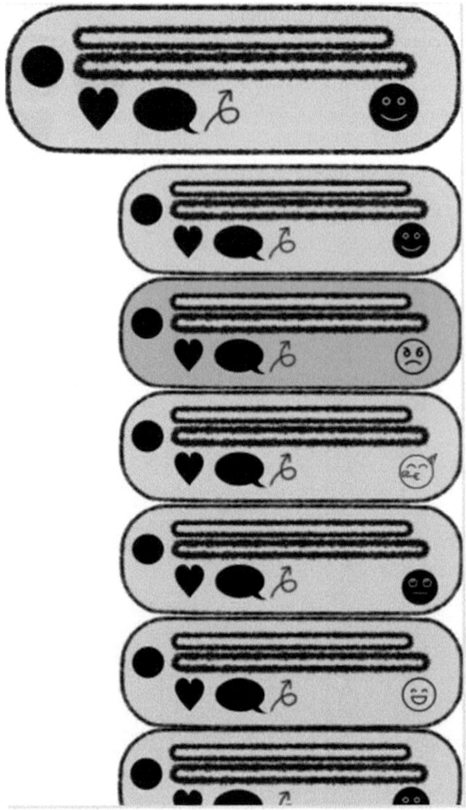

Fig. 3.4 Angry comment reintegrated

Fig. 3.5 Comment queuing on a thread

thresholds for anger or fear might be temporarily increased to lessen the number of queued comments. On the other hand, during calmer times, the thresholds may be marginally reduced to enable stricter moderation and avert possible intensification of strong emotions. The active/non-active stage of the conversation is distinguished based on continuous analysis of time intervals between comments, overall comment volume, and user engagement patterns. This ensures the system can adapt to activity levels, such as high-intensity discussions or subdued phases.

The threshold values also adjust to the conversation's context and evolving emotional distribution. A sliding window approach is employed, focusing on the most recent comments to capture real-time emotional states and prevent outdated emotions from distorting the tone. In our experiments, we used 100 recent comments. Positive emotions, like joy or love, are given higher weight through weighted allowances, promoting constructive interactions while moderating negative emotions like anger or fear. To minimise comment suspension and maintain emotional balance, the system prioritises underrepresented emotions in the conversation. This ensures that diverse emotional expressions are integrated, preventing the dominance of a single emotional tone while promoting a more balanced, inclusive dialogue.

By incorporating a reflective pause and offering users the opportunity to refine their contributions, this system promotes more deliberate and emotionally balanced interactions, while also discouraging impulsive comments.

For example, in a particularly active Twitter/Reddit thread discussing pandemic policies, the thresholds for anger and fear would be temporarily elevated due to the influx of comments expressing intense emotions. This adjustment will help reduce disruptions in the conversation, enabling comments with moderate emotional weight to be included while ensuring that extreme outliers to be managed and queued appropriately.

3.3.5 Experimental Setup

As shown in Fig. 3.1, following the processing of the collected data, each conversation was converted into a directed acyclic conversation graph. Subsequently, every comment in the graph was assigned an emotion using the NRC lexicon, along with the intensity of the emotion contained in the comment's content. The influence of each comment was computed based on a combination of its emotional intensity, distance from the root node, PageRank, and number of replies. This score was utilised to update the emotional impact of the conversation. The root node maintained an emotion board that tracked the percentage influence of each emotion on the conversation, with this board being dynamically updated as new comments were added to the graph. The framework was assessed by comparing its performance to the toxicity scores generated by the Google Perspective API (2021). Our analysis focused on mitigating hate speech and polarisation by deactivating influential nodes or restricting responses once they reach the toxicity threshold. Our findings revealed that, while the Perspective API reduced hate speech by 7%, the eImpact framework achieved a 10% reduction. This confirms the effectiveness of the proposed framework in identifying and regulating post-toxicity while considering its subjectivity. Moreover, it can complement online content moderation efforts by providing insights into the sources of toxicity in conversations, taking both content and context into consideration.

To simulate the evolving nature of online conversations, dynamic thresholds were introduced for detecting toxicity. These thresholds decreased as more comments were processed, enabling a flexible and adaptive assessment of emotional influence. In the scenario of queuing, comments that exceeded the toxic thresholds on the root node's emotion board were placed in a queue. These held comments were reassessed after each new comment addition, and those meeting the threshold were subsequently added to the graph.

3.4 Results and Discussion

In this study, we assessed the efficacy of the eImpact framework and the comment queuing mechanism in managing emotional dynamics within online conversations represented as graphs, with a specific emphasis on moderating toxic comments. We compared two different approaches: one without a queuing mechanism, where comments were added to the graph immediately, and one with a queue, where comments were temporarily held and re-evaluated against emotional thresholds before being incorporated into the conversation.

Our analysis of emotional propagation in online conversations, using data from Twitter and Reddit, revealed that highly emotional content tends to receive greater user engagement. We observed that tweets featuring emotive language, such as expressions of love, often triggered intense, anger-fuelled reactions, leading to heightened emotional polarisation. This polarisation effect was pronounced when the same group of users repeatedly interacted with the post, resulting in the amplification of anger and the suppression of other emotions like joy. Our findings are consistent with existing research indicating that deeply rooted radical viewpoints are more likely to manifest as toxicity when anger dominates online discourse. Figure 3.6 shows the average distribution of emotions and the number of unique users involved in influential node reply trees for each emotion represented by the influential node.

We evaluated the eImpact framework, which is designed to mitigate emotional escalation, alongside Google's Perspective API for mitigating toxic content. The results demonstrated that eImpact led to a 10% reduction in hate speech and polarising content, compared to a 7% decrease with the Perspective API. This outcome emphasises the proposed framework's effectiveness in addressing not only the content but also the broader context and emotional dynamics within conversations. By prompting users to consider the emotional impact of their comments, eImpact promotes self-regulation and reduces toxic behaviour over time. Table 3.1 compares the proposed framework to the Perspective API in terms of identifying influential nodes and the possibility of toxicity reduction.

We then implemented the comment queuing mechanism to moderate emotional intensity in real time. The queuing system managed comment posting by assessing their emotional impact before allowing them into the conversation. Analysis of the held durations revealed that comments were typically queued for 40–55 s, with an average hold time of 47 s, providing ample time for self-reflection. Figure 3.7 shows

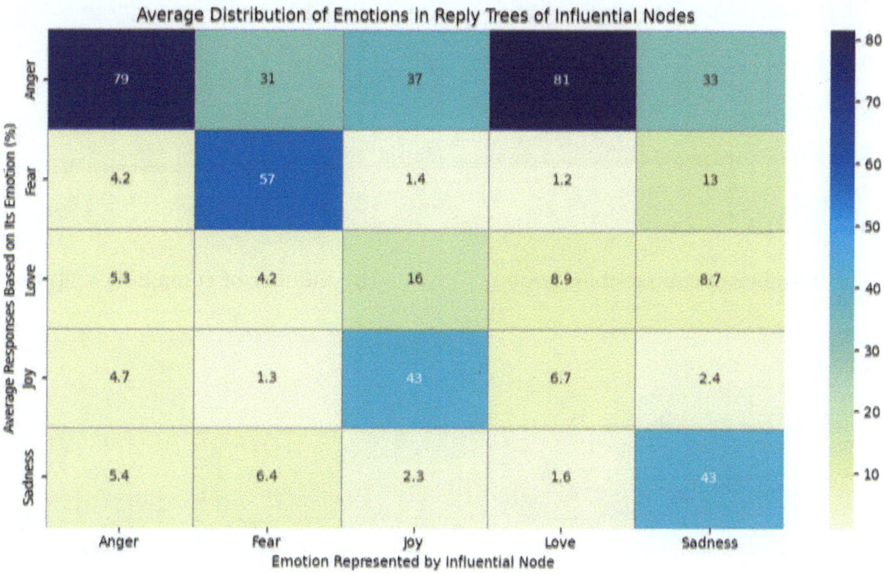

Fig. 3.6 Average distribution of emotions and the number of unique users in the reply trees of influential nodes (Verma et al. 2023)

Table 3.1 Comparison of proposed framework (Verma et al. 2023) against Perspective API (2021)

Utilised model	Identifying influential nodes	Toxic node estimated toxicity	
		Detection	Reduction
eImpact Framework	Based on impact score, taking into account the tweet text as well as its connectivity	1–4%	10%
Perspective API (2021)	Based on toxicity, taking into account only the tweet text	1–2%	7%

a histogram that illustrates the time distribution of comments held in the queue before being processed. The x-axis shows the duration of being held in seconds, while the y-axis represents the frequency of comments. As shown in Fig. 3.7, significant concentration of comments with short-held durations indicates that the queue efficiently reintegrated comments into the flow. The distribution indicates that the majority of comments were held for brief periods, suggesting that the system effectively moderated emotions without causing substantial delays. Longer-held durations point to comments that posed greater emotional challenges, requiring extended holding times which is required to ensure the stability of the emotion board.

When comparing conversations with and without the queuing system, we observed that the emotion board maintained a more balanced distribution of emotions when the queue was employed. Dominant emotions such as anger and fear, which were

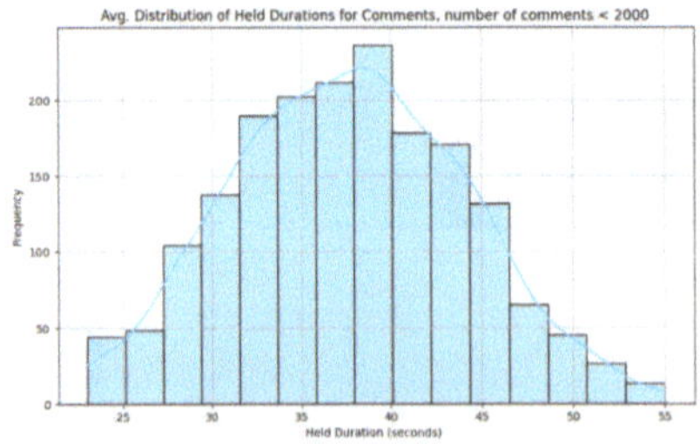

(a) Number of comments < 2000

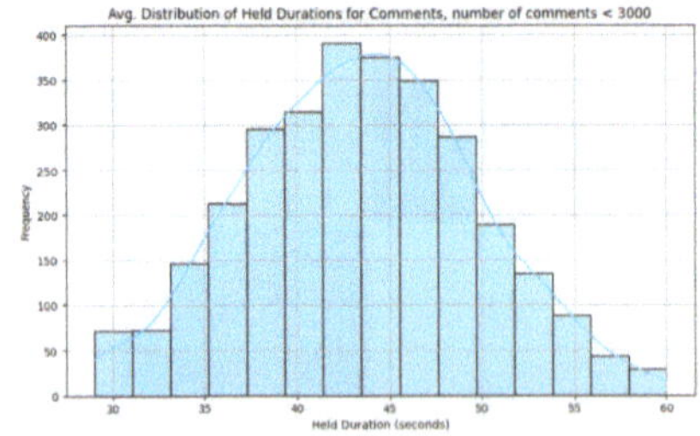

(b) Number of comments < 3000

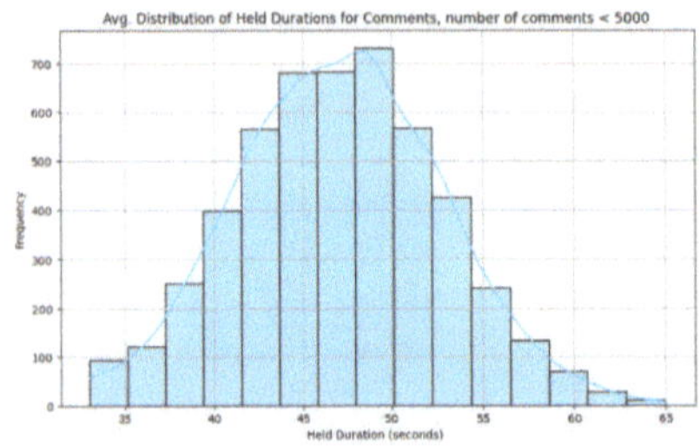

(c) Number of comments < 5000

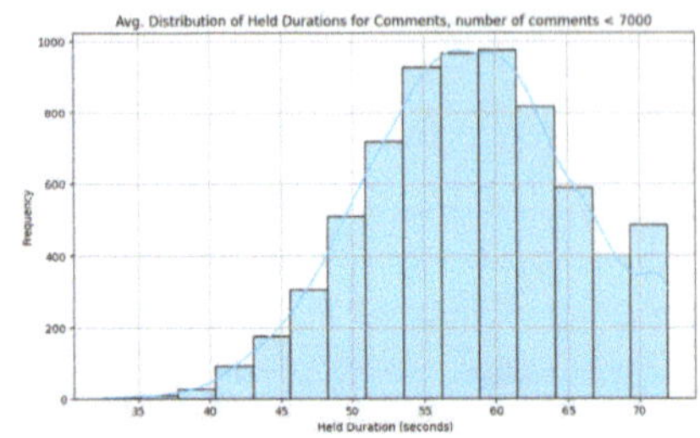

(d) Number of comments < 7000

Fig. 3.7 Average distribution of held durations for comments when using the proposed queue approach

prevalent without the queue, were significantly reduced. The gradual integration of comments into the conversation helped slow down emotional spikes, particularly in cases where comments could have rapidly escalated negative emotions. As a result, the conversation took on a more moderated tone. Figure 3.8 showcases bar graphs that compare the final emotional composition of the conversation graph under the two experimental conditions—without queuing and with queuing. The y-axis shows the percentage contribution of each emotion to the overall emotional state of the root node. These graphs showcase the influence of the queuing mechanism on emotional dynamics. In the "Without Queue" scenario, the conversation is predominantly characterised by negative emotions such as Anger and Fear, while the "With Queue" scenario presents a more balanced emotional state. When the queue is used, emotions are always within threshold limits and not dominated by specific emotions. The findings suggest that in the absence of queuing, negative emotions tend to accumulate and dominate the conversation. Conversely, the queuing mechanism effectively moderates emotional impact, preventing negative emotions from overpowering the conversation and resulting in a more balanced emotional distribution.

The dynamic thresholds, sliding windows, and weighted allowances in the queue mechanism contributed to sustaining the flow of the conversation. By adapting thresholds in response to the context and activity level of the conversation, the system was able to evaluate comments in real time. The sliding window, which concentrated on the most recent 100 comments, ensured that only the most recent emotional trends

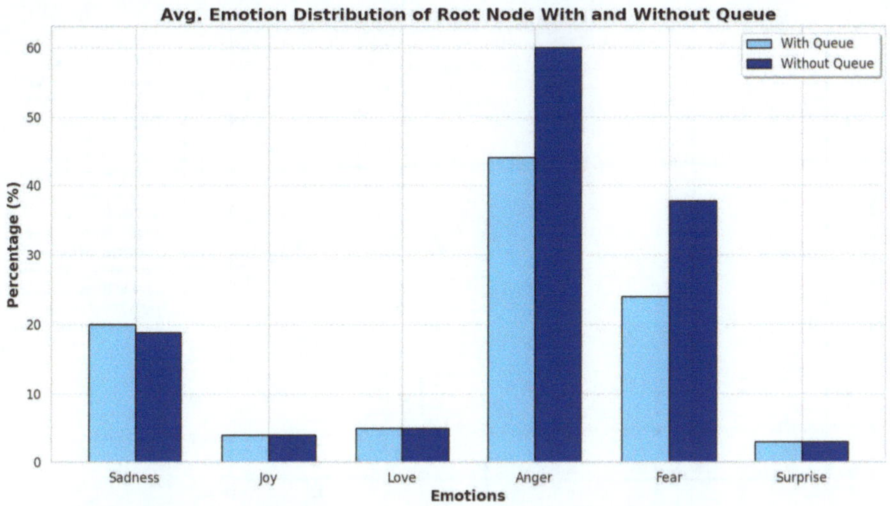

Fig. 3.8 Average emotion distribution of root node with and without queue

influenced the conversation, thus preventing outdated emotions from influencing the overall tone. Moreover, by prioritising positive emotions and assigning higher priority to underrepresented emotions in the queue, the system facilitated an emotionally balanced conversation. Figure 3.9 shows a line graph illustrating the changing emotional impact over time for each emotion within the conversation graph when the queuing mechanism is implemented. The x-axis denotes time, while the y-axis represents the cumulative emotional impact. This graph offers a visualisation of how emotions evolve as the conversation unfolds and comments are added sequentially. It provides insight into how the queuing mechanism influences the flow of specific emotions. The graph indicates that negative emotions experience surges but are mitigated at various points, suggesting that the queuing mechanism intervenes to prevent them from dominating the conversation. On the other hand, positive emotions such as Joy and Love intensify over time, indicating that the system facilitates a more positive emotional trajectory.

Overall, when the queue was employed, we observed an average reduction of 15% in the spread of anger and fear compared to conversations where comments were posted immediately. This reduction highlights the potential of the queuing system to encourage healthier and less emotionally charged conversations by allowing users the time to reconsider their comments. Furthermore, only 4% of comments were held for review, demonstrating the system's effectiveness in maintaining a smooth flow of conversation while managing emotional escalation.

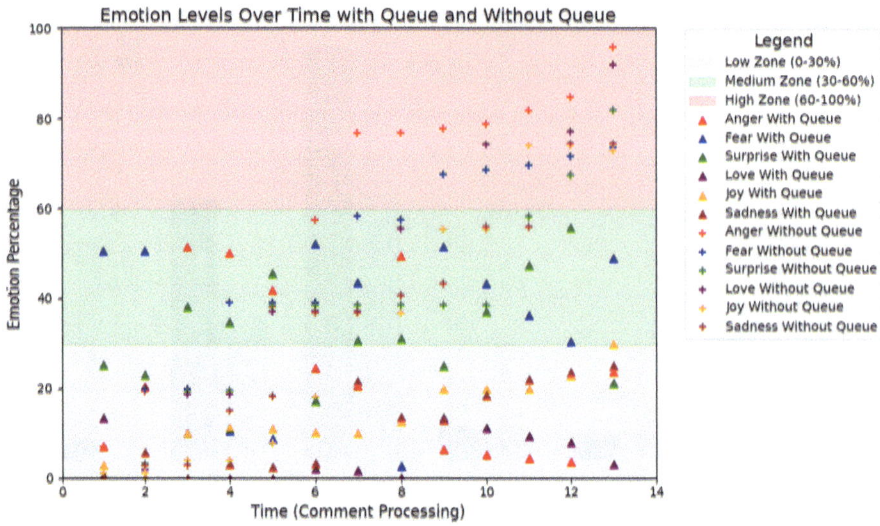

Fig. 3.9 Average emotion levels in the conversation when using the queue

3.4.1 Conclusion

This paper presents an innovative framework that combines comment queuing with adaptive thresholds to decrease online toxicity and facilitate emotion regulation in digital conversations. By holding potentially harmful comments in a queue and evaluating their impact before publishing, the framework promotes self-reflection and mitigates emotional spikes. Our experiments showed a 15% reduction in negative emotions such as anger and fear compared to unmoderated conversations.

This framework makes a valuable contribution to the field of Digital Emotion Regulation (DER) by encouraging healthier and more responsible online discourse without compromising user engagement. Future research can build upon this work by implementing adaptive mechanisms for different types of content and exploring real-time emotional feedback through user interviews and surveys.

Acknowledgements During the drafting of this paper, Grammarly (2024) was used to check and enhance the grammar and writing style of this document. We are grateful to Amity University for accepting our work as a keynote paper.

References

Barry E, Jameel S, Raza H (2021) Emojional: Emoji embeddings. In: UK workshop on computational intelligence. Springer, pp 312–324

Buckels EE, Trapnell PD, Paulhus DL (2014) Trolls just want to have fun. Person Indiv Diff 67:97–102

Chandrasekharan E, Jhaver S, Bruckman A, Gilbert E (2022) Quarantined! Examining the effects of a community-wide moderation intervention on reddit. ACM Trans Comput Human Interact 29(4):1–26
Chandrasekharan E, Pavalanathan U, Srinivasan A, Glynn A, Eisenstein J, Gilbert E (2017) You can't stay here: the efficacy of reddit's 2015 ban examined through hate speech. Proc ACM Human Comput Interact 1(CSCW):1–22
Cheng J, Bernstein M, Danescu-Niculescu-Mizil C, Leskovec J (2017) Anyone can become a troll: causes of trolling behavior in online discussions. In: Proceedings of the 2017 ACM conference on computer supported cooperative work and social computing, pp 1217–1230
Frischlich L, Schatto-Eckrodt T, Boberg S, Wintterlin F (2021) Roots of incivility: how personality, media use, and online experiences shape uncivil participation. Media Commun 9(1):195–208
Gillespie T (2018) Custodians of the internet: platforms, content moderation, and the hidden decisions that shape social media. Yale University Press
Gillespie T (2020) Content moderation, AI, and the question of scale. Big Data Soc 7(2):2053951720943234
Goel S, Anderson A, Hofman J, Watts DJ (2016) The structural virality of online diffusion. Manage Sci 62(1):180–196
Gongane VU, Munot MV, Anuse AD (2022) Detection and moderation of detrimental content on social media platforms: current status and future directions. Soc Netw Anal Min 12(1):129
Google: Perspective API (2021) https://www.perspectiveapi.com/. Accessed 19 Dec 2022
Gorwa R, Binns R, Katzenbach C (2020) Algorithmic content moderation: technical and political challenges in the automation of platform governance. Big Data Soc 7(1):2053951719897945
Grammarly (2024) Grammarly. https://www.grammarly.com. Accessed 20 Jun 2024
Gross JJ (2008) Emotion regulation. Handbook Emot 3(3):497–513
Gross JJ (2015) Emotion regulation: current status and future prospects. Psychol Inq 26(1):1–26
Gross JJ (2014) Emotion regulation: conceptual and empirical foundations
Grossmann I, Kross E (2014) Exploring Solomon's paradox: self-distancing eliminates the self-other asymmetry in wise reasoning about close relationships in younger and older adults. Psychol Sci 25(8):1571–1580
Hadwin A, Järvelä S, Miller M (2017) Self-regulation, co-regulation, and shared regulation in collaborative learning environments. In: Handbook of self-regulation of learning and performance. Routledge, pp 83–106
Hadwin AF, Davis SK, Bakhtiar A, Winne PH (2019) Academic challenges as opportunities to learn to self-regulate learning. In: Problem solving for teaching and learning, pp 34–47
Herwig U, Kaffenberger T, Jäncke L, Brühl AB (2010) Self-related awareness and emotion regulation. Neuroimage 50(2):734–741
Hossain E, Wadley G, Berthouze N, Cox A (2022) Motivational and situational aspects of active and passive social media breaks may explain the difference between recovery and procrastination. In: CHI conference on human factors in computing systems extended abstracts, pp 1–8
Jane EA (2020) Online abuse and harassment. In: The international encyclopedia of gender, media, and communication, vol 116
Kiskola J, Olsson T, Väätäjä H, Syrjämäki HA, Rantasila A, Isokoski P, Ilves M, Surakka V (2021) Applying critical voice in design of user interfaces for supporting self-reflection and emotion regulation in online news commenting. In: Proceedings of the 2021 CHI conference on human factors in computing systems, pp 1–13
Kiskola J, Olsson T, Syrjämäki AH, Rantasila A, Ilves M, Isokoski P, Surakka V (2022) Online survey on novel designs for supporting self-reflection and emotion regulation in online news commenting. In: Proceedings of the 25th international academic Mindtrek conference, pp 278–312
Kou Y, Gui X (2020) Emotion regulation in esports gaming: a qualitative study of league of legends. Proc ACM Human Comput Interact 4(CSCW2):1–25
Kumar S, Cheng J, Leskovec J (2017) Antisocial behavior on the web: characterization and detection. In: Proceedings of the 26th international conference on world wide web companion, pp 947–950

Lukoff K, Yu C, Kientz J, Hiniker A (2018) What makes smartphone use meaningful or meaningless? Proc ACM Interact Mobile Wear Ubiq Technol 2(1):1–26
Luo RND, Yang XJ (2021) Behavioral and physiological signals-based deep multimodal approach for mobile emotion recognition. IEEE Trans Affect Comput
Maarouf A, Pröllochs N, Feuerriegel S (2022) The virality of hate speech on social media. arXiv preprint arXiv:2210.13770
Manikonda L, Beigi G, Liu H, Kambhampati S (2018) Twitter for sparking a movement, reddit for sharing the moment:# metoo through the lens of social media. arXiv preprint arXiv:1803.08022
McRae K, Gross JJ (2020) Emotion regulation. Emotion 20(1):1
Mohammad, SM, Turney PD (2013) Nrc emotion lexicon, vol 2. National Research Council, Canada, p 234
Ochsner KN, Gross JJ (2005) The cognitive control of emotion. Trends Cogn Sci 9(5):242–249
Park JS, Seering J, Bernstein MS (2022) Measuring the prevalence of anti-social behavior in online communities. Proc ACM Human Comput Interact 6(CSCW2):1–29
Phillips W (2015) This is why we can't have nice things: mapping the relationship between online trolling and mainstream culture. MIT Press
Ruensuk M, Cheon E, Hong H, Oakley I (2020) How do you feel online: exploiting smartphone sensors to detect transitory emotions during social media use. Proc ACM Interact Mobile Wear Ubiq Technol 4(4):1–32
Saveski M, Roy B, Roy D (2021) The structure of toxic conversations on twitter. In: Proceedings of the web conference, pp 1086–1097
Shen K, Cox A (2020) Video games as a tool for digital emotion regulation. HCI-E MSc Final Project Report 2020
Slovak P, Antle AN, Theofanopoulou N, Roquet CD, Gross JJ, Isbister K (2022) Designing for emotion regulation interventions: an agenda for HCI theory and research. arXiv preprint arXiv: 2204.00118
Smith W, Wadley G, Webber S, Tag B, Kostakos V, Koval P, Gross JJ (2022) Digital emotion regulation in everyday life. In: CHI conference on human factors in computing systems, pp 1–15
Solovev K, Pröllochs N (2022) Moral emotions shape the virality of covid-19 misinformation on social media. In: Proceedings of the ACM web conference 2022, pp 3706–3717
Thomas K, Kelley PG, Consolvo S, Samermit P, Bursztein E (2022) "It's common and a part of being a content creator": understanding how creators experience and cope with hate and harassment online. In: CHI conference on human factors in computing systems, pp 1–15
Torre JB, Lieberman MD (2018) Putting feelings into words: Affect labeling as implicit emotion regulation. Emot Rev 10(2):116–124
Trujillo A, Cresci S (2022) Make reddit great again: assessing community effects of moderation interventions on r/the_donald. Proc ACM Human Comput Interact 6(CSCW2):1–28
Verma A, Islam S, Moghaddam V, Anwar A (2023) Encouraging emotion regulation in social media conversations through self-reflection. In: 2023 IEEE engineering informatics, pp 1–8. https://doi.org/10.1109/IEEECONF58110.2023.10520471
Wadley G, Smith W, Koval P, Gross JJ (2020) Digital emotion regulation. Curr Dir Psychol Sci 29(4):412–418
Yin D, Xue Z, Hong L, Davison BD, Kontostathis A, Edwards L et al (2009) Detection of harassment on web 2.0. Proc Content Anal WEB 2:1–7

Chapter 4
Effect of Multiwalled Carbon Nanotubes on the Compressive Strength and Microstructural Property of Rubberised Concrete Using RSM Modelling

Vasudevan Arveendh, Naraindas Bheel, Abhijeet Vidyadhar Baikerikar, and Bashar S. Mohammed

Abstract The incorporation of crumb rubber (CR) obtained from waste tyres as a substitute for sand in concrete, resulting in the making of rubberised concrete, has led to a variety of environmental advantages and improved characteristics of the concrete. Although there have been certain enhancements, the decreased strength of rubberised concrete has restricted its wide application. This work presents an inventive technique to address this issue, integrating multiwalled carbon nanotubes (MWCNTs) to develop rubberised concrete with a compressive strength (CS) comparable to conventional mixes. Thirteen combinations were created adopting the response surface methodology (RSM). These mixtures consisted of varying quantities of MWCNTs, fluctuating from 0.05 to 0.08%, and CR substituted for 1–5% of the sand volume. The purpose of these mixtures was to determine the CS and microstructural characteristics. The finding has displayed that the CS of CR concrete is observed by 39.90 MPa, 45.49 MPa, and 57.60 MPa at 0.05% of MWCNTs and 1% of CR after 7, 14, and 28 days, respectively. Besides, the microstructural properties are getting improved while the content of MWCNTs rises in CR concrete at 28 days. It is concluded that the use of 0.05% of MWCNT along with 1–3% of CR in concrete is offering the best outcomes for the construction applications.

V. Arveendh · N. Bheel · A. V. Baikerikar · B. S. Mohammed (✉)
Department of Civil and Environmental Engineering, Universiti Teknologi PETRONAS, Bandar Seri Iskandar, Perak, Malaysia
e-mail: bashar.mohammed@utp.edu.my

V. Arveendh
e-mail: arveendh_18002518@utp.edu.my

N. Bheel
e-mail: naraindas_20001014@utp.edu.my

A. V. Baikerikar
e-mail: abhijeet_22012282@utp.edu.my

B. Shukla et al. (eds.), *Sustainable Technologies for Transforming Businesses and Societies*, Transactions on Computer Systems and Networks,
https://doi.org/10.1007/978-981-96-9471-6_4

Keywords MWCNTs · Crumb rubber · Compressive strength · Microstructural property · RSM modelling

4.1 Introduction

Rubber tyres provide an essential backbone of road transportation. Global manufacturers produce approximately 1.2 billion rubber tyres annually. To meet the demands for rapid transportation, a significant amount of rubber tyres is generated after they have reached the end of their useful lifespan. It is projected that around 5 billion tyres will be discarded by 2030 after they have served their expected lifetime (Eltayeb et al. 2021). Based on statistics provided by the International Rubber Research Group, global rubber production in 2018 amounted to 29.39 million tonnes. In 2019, this statistic saw a 2.5% rise, attaining a total of 30.12 million tonnes (Li et al. 2019). The disposal of this huge quantity of used rubber tyres presents significant challenges. The main methods of solid waste management include incineration, landfilling, and decomposition. Nevertheless, landfills occupy important land areas, and rubber waste has a slow degradation process that poses a continuous environmental hazard (Chaturvedy et al. 2023). The combustion of crumb rubber (CR) emits toxic fumes, resulting in legislative bans in several nations owing to its adverse impact on people, wildlife, and plants (Mohammed and Adamu 2018; Mohammed et al. 2018a, b, cs; Adamu et al. 2018; Azmi and Nasional 2008). Rubber reprocessing based on pyrolysis, which aims to produce carbon black powder, has the potential to contaminate the environment and produce a more expensive and inferior product than petroleum-derived carbon black. As a result, the incorporation of waste rubber into building materials demonstrates its efficacy. It decreases the demand for natural aggregates in construction, thereby preserving land and resolving rubber waste management problems (Chaturvedy and Pandey 2022).

Crumb rubber is obtained from the outer surface of used tyres. It is utilised as a substitution for sand in concrete, leading to a kind of concrete known as CR concrete or rubberised concrete (Mohammed et al. 2017; Sabapathy et al. 2020; Al-Fakih et al. 2021; Khed et al. 2020; Adamu et al. 2018, 2020). This unique form of concrete offers numerous advantages than that of standard concrete, such as decreased density (Demir et al. 2015; Mohammed et al. 2016), enhanced ductility (Shu and Huang 2014), boosted plastic capability (Mohammed et al. 2011), better resistance to impact (Ganjian et al. 2009), greater durability (Mohammed 2010), higher resistance to penetration (Bravo and Brito 2012), higher electrical resistivity (Onuaguluchi and Panesar 2014), and lesser thermal conductivity (Mohammed et al. 2012). Furthermore, rubberised concrete is renowned for its exceptional capacity to dissipate energy, its long-lasting nature, and its high dampening coefficient (Youssf et al. 2014). However, rubberised concrete does have certain disadvantages. One notable issue is that as the fraction of CR substituting sand increases, there is a drop in CS, tensile, and flexural strengths. The reduction in strength might be attributed to the insufficient adhesion between CR and the PC mixture component, which is caused by the hydrophobic properties of CR induced throughout tire production using zinc stearate (Youssf et al. 2014). The hydrophobic nature of CR creates an impermeable, non-polar barrier, resulting in an augmented ITZ between the CR and PC mix. The existence of a more compact area weakens the strength of the connection, intensifies stress, and eventually leads to the creation of tiny cracks, which break prematurely (Mohammed et al. 2012; Sadek and El-Attar 2015; Li et al. 2016; Thomas et al. 2016).

In order to enhance the practical application and widespread adoption of CR concrete in the building sector, significant investments have been directed into augmenting its strength. Scientists have investigated several techniques to enhance the bonding of CR particles and PC mix by treating the surface of the CR (Al-Fakih et al. 2020; Abdulkadir et al. 2020; Adamu et al. 2018, 2019; Hong et al. 2020). Rostami et al. (2000) performed research to assess the efficacy of several techniques in treating different surfaces, including water, a carbon tetrachloride solution, and a coating additive cleanser. The outcomes showed that the compressive strength (CS) augmented by up to 57% compared to the standard combination. Segre and Joekes (2000) obtained favourable results by utilising a NaOH solution to clean the surfaces of CR, thereby improving adhesion and ultimately increasing strength. Albano et al. (2005) did not see any significant improvements in mechanical strength after treatment CR with NaOH and silane bonding substances. Chou et al. (2010) investigated the process of partially oxidising CR in order to provide hydrophilic groups, which enhance the binding with cement matrix. However, this method comes with higher expenses. Pelisser et al. (2011) conducted an experiment in which they combined NaOH treatment with silica fume in rubberised concrete. They observed that this treatment led to a less significant reduction in strength compared to rubberised concrete without any treatment. Huang et al. (2013) and Dong et al. (2013) focused experiments on coating the surface using a two-phase approach and found that the combination of both phases was more efficient in boosting the strength of rubberised concrete. Meddah et al. (2014) conducted an experiment in which they used NaOH pretreatment and added fine aggregate to raise the surface roughness of CR. However, this procedure did not result in a substantial enhancement in the strength of concrete. In research conducted by Li et al. (2016), coupling agents made from silane and carboxylate styrene-butadiene rubber latex were utilised for treating old tire fibres. They observed improvements in the composites' CS and flexural strength. The use of CR treatment procedures may not consistently produce rubberised concrete with the same strength as conventional concrete, contrary to previous studies by Dong et al. (2013) and Shu and Huang (2014). This diminishes its use in buildings due to its inferior strength compared to conventional concrete. The developments in nanotechnology have great potential for nanoscale material production, allowing for precise material control at the atomic level. Nanomaterials have enhanced performance compared to their larger-scale counterparts owing to their smaller dimensions (Faried et al. 2021; Nik et al. 2010; Bahari et al. 2016; Mousavi et al. 2021). Multi-walled carbon nanotubes (MWCNTs) have become a prominent nanomaterial in the field of construction, fulfilling the requirement and addressing the issues of rubberised concrete.

In addition, the incorporation of low-dosage MWCNTs may significantly enhance the hardened state of the substance and could have a direct influence on its composition in specific ways (Wang et al. 2022; Lee et al. 2010; Sldozian et al. 2019). Subsequently, lots of experts have shown interest in MWCNTs due to their extraordinary properties (Zhang et al. 2023; Iijishima 1991). MWCNTs, or multiwalled carbon nanotubes, are cylindrical nanostructures composed of numerous layers of graphene sheets coiled around a central axis (Bheel et al. 2024, 2023a). The diameters associated with these cylindrical constructions may range from 2 to 100 nm, based on the amount of graphene layers or walls. The distance between these layers is approximately 0.34 nm, which is similar to the gap between layers in graphite. MWCNTs (Li et al. 2020; Nam et al. 2017) possess exceptional mechanical, electrical, and thermal conductive properties. Eftekhari and Mohammadi (2016) showed via simulated testing that the use of MWCNTs in concrete could boost its strength. Carriço et al. (2018) found that concrete with a high concentration of MWCNTs and a small aspect ratio demonstrated enhanced durability. Konsta-Gdoutos et al. (2010) observed that the inclusion of carbon nanotubes (CNTs) significantly enhanced the strength of the C–S–H bonds in composites. This study shows that carbon nanotubes (CNT) have the potential to provide a more reliable and appropriate solution for hydration characteristics (Zhang et al. 2023). The results of Liew et al. (2016) indicate that carbon nanotubes (CNT) alter the microscopic structure of the concrete matrix and reduce the total porosity. The main reasons that contribute to this phenomenon are the three roles performed by MWCNTs, which are gap filling, bridging, and nuclei production. Therefore, this study is conducted to investigate the compressive strength, and microstructural property of CR concrete blended with various contents of MWCNTs by applying response surface methodology (RSM) modelling to optimise the outcome and input parameters.

4.2 Materials and Methods

4.2.1 Materials

The Portland cement (PC) was applied in the investigational assessment as binding agent which meets the criteria of ASTM C150 (ASTM 2015) and its composition as shown in Table 4.1. However, the multiwalled carbon nanotubes (MWCNTs) were utilised as additional material by the mass of PC in research work. Besides, the sand is sourced from the riverbank and is in its natural state, and then it passes through a 4.75 mm sieve. After sieving, it was employed as fine aggregate (FA) in the research work and its specific gravity is 2.64. Moreover, the crushed stone was applied as coarse aggregate (CA) in the investigational study having a 20 mm in size and its specific gravity is about 2.66. In addition, the crumb rubber (CR) materials were carefully varied from 0 to 5% and applied as a partial substitute for sand based on volume. The material was sieved through a 1.18 mm sieve, and the specific gravity of the CR is 0.92. Furthermore, the drinkable water was utilised in this experimental work for curing and mixing of concrete.

4.2.2 Mix Proportions Generated by RSM

The purpose of this study was to apply the RSM to investigate the impact of two independent parameters, namely CR and MWCNTs, on the investigation outcome. The parameters were tested at three substitution ratios of 1–5% of sand replaced with CR and 0.05–0.08% of MWCNTs by the mass of PC, correspondingly. A total of 13 experiment runs were created utilising the RSM technique. Table 4.2 displays the mixtures with different mixtures and degrees of input components, along with five randomised repetitions for each parameter. The duplicated mixtures were created to assure the experiment's efficacy and mitigate any variations (Mohammed et al. 2020). The RSM quantifies the impact of the relationship between the input elements on the output. The evaluated parameters included the CS at 7, 14, and 28 days. Furthermore, the amounts of ingredients needed to generate the experimental specimens are demonstrated in Table 4.2.

4.2.3 Sample Preparation and Testing Methods

The specimens were fabricated using thoroughly blended CR concrete, which was manufactured in accordance with the standards outlined in BS 1881: Part 125:1986. The sand, CA, and CR were thoroughly blended for a duration of 2 min in a concrete mixer. Subsequently, water was combined with MWCNTs and SP to effectively disseminate the MWCNTs via sonication. In order to counterbalance the decrease in fluidity and achieve a manageable mixture, the addition of SP was implemented in mixtures containing higher levels of MWCNTs. The fresh mix was divided into several specimens for the purpose of conducting compressive strength (CS) testing. The CS measurement was conducted in accordance with the standards stated in BS EN 12390-3 (BS EN 2009), with curing durations of 7, 14, and 28 days. The CS of the combination for a certain curing period is determined by calculating the mean of three outcomes.

Table 4.1 Chemical composition of PC

Materials	Compound (%)									Specific Gravity	Blaine fineness (m^2/kg)	Loss on ignition
	SiO_2	Fe_2O_3	Al_2O_3	CaO	MnO	Na_2O	MgO	T_2O	K_2O			
PC	20.76	3.35	5.54	61.4	–	0.19	2.48	–	0.78	3.15	325	2.20

Table 4.2 Mix proportions of CR concrete blended with MWCNTs

Mix ID	Ingredients (%)	Amount of ingredients utilised in CR concrete (kg/m^3)				
	MWCNTs	CR	PC	CA	Sand	Water
M0	0.00	0	575	1425	662	202
M1	0.065	3	575	1425	662	202
M2	0.08	5	575	1425	662	202
M3	0.065	1	575	1425	662	202
M4	0.065	5	575	1425	662	202
M5	0.065	3	575	1425	662	202
M6	0.08	3	575	1425	662	202
M7	0.05	1	575	1425	662	202
M8	0.05	5	575	1425	662	202
M9	0.08	1	575	1425	662	202
M10	0.065	3	575	1425	662	202
M11	0.065	3	575	1425	662	202
M12	0.05	3	575	1425	662	202
M13	0.065	3	575	1425	662	202

4.3 Results and Discussions

4.3.1 Compressive Strength (CS)

Figure 4.1 displays the CS outcomes of CR concrete blended with various contents of MWCNTs at 7, 14, and 28 days correspondingly. The maximum CS was noted by 39.90, 45.49, and 57.60 MPa at 0.05% of MWCNTs and 1% of CR as sand replacement together while the lowest CS was recorded by 29.80, 33.80, and 42.97 MPa at 0.08% of MWCNTs along with 5% of sand replaced with CR by volume after 7, 14, and 28 days consistently. Figure 4.1 demonstrates a clear correlation between the rise in CR percentage in concrete and the decrease in CS. The primary cause for the decline in CS as the CR concentration increases is the inadequate connection between hardened PC mixtures and the CR during the point of interaction, as shown by Nazari et al. (2010). The inadequate bonding occurred due to the hydrophobic characteristics of CR components, which lead them to repel water when mixed (Abdulkadir et al. 2021). Furthermore, the CR leads to enhanced pores in composites, thus diminishing their strength. Moreover, the CR's soft texture and low elastic modulus, in comparison with the sand particles, resulted in reduced concrete strength. This is because the CR acted as weak spots throughout the composite, as shown in earlier studies (Kundan and Sharma 2020; Assaggaf et al. 2021; Fauzan et al. 2021; Abdulkadir et al. 2021). This finding lines up with other study that has verified the detrimental impact of additional CR on the CS of composites (Zhang et al. 2015; Alaloul et al. 2020). Furthermore, the MWCNTs have the capacity to augment the CS of CR concrete

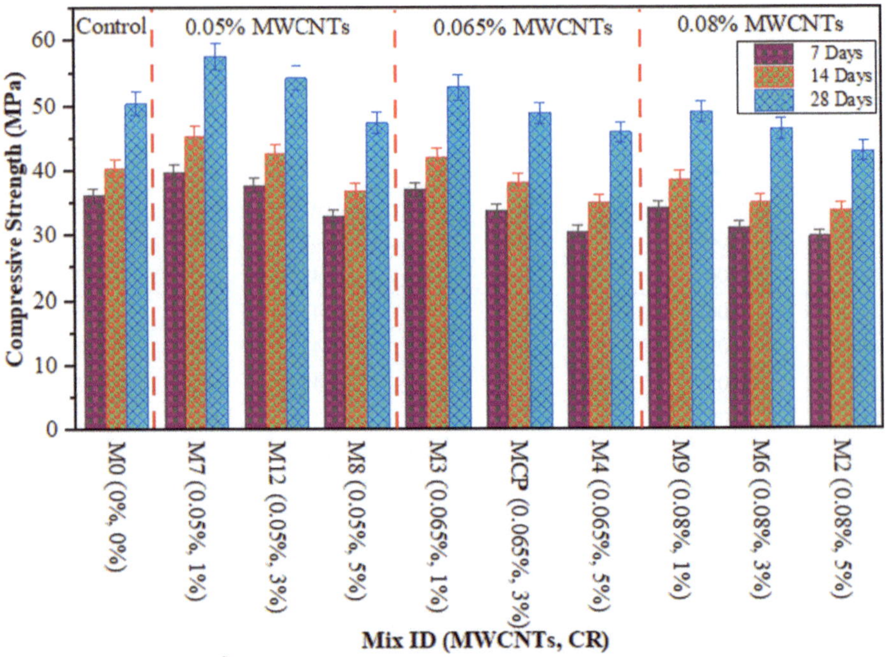

Fig. 4.1 CS of CR concrete blended with MWCNTs

when included at a concentration of up to 0.05% relative to the weight of PC, as illustrated in Fig. 4.1. This enhancement in CS of CR concrete is associated with the pore filling effect of MWCNTs which seals the free spaces in CR concrete and makes it stiffer CR concrete that results in improving CS (Lv et al. 2013; Mohammed et al. 2015; Pan et al. 2015; Gong et al. 2014). Conversely, the improvement in the CS of CR concrete may be outstanding dispersion of MWCNTs (Khan et al. 2023). For further addition of MWCNTs in CR concrete, the CS starts reducing. This drop in CS of CR concrete is correlated to the inadequate MWCNTs dispersion in CR concrete matrix which negatively affect on CS of CR concrete. The similar type of findings was aligned to the prior research (Khan et al. 2023; Zheng et al. 2021).

4.3.2 Mercury Intrusion Porosimetry (MIP) Analysis

A MIP analysis was performed on a specific set of samples to examine the properties of the pores and assess the way distinct inputs affect the structure and distribution of the pores. To understand the distribution of pore sizes, the most effective method is to analyse the pore size distribution curve (PSDC) revealed in Fig. 4.2a–f. Frequently, the solid by-products generated by the hydration procedure have a smaller volume compared to the combined volume of the water and cement before any hydration occurs. Capillary holes fill the remaining volume discrepancy. The pores have a nominal diameter dimension that ranges from more than one millimetre to less than 100 nm. Nevertheless, making the hydrated results more solid would fill up all the capillary voids, reducing their volume and breaking down the connections between them. The solid by-products of the process of hydrate identification, identified as gel, have holes that are formed as a result of water evaporation. These spaces are referred to as gel pores. The size of these holes is lower than that of capillary pores, with a nominal diameter of less than 3 nm. Nevertheless, gel pores constitute around 28% of the overall gel volume (Mohammed et al. 2016; Neville 2011).

Thus, the limited dimensions of gel pores restrict the physical filling to just MWCNTs. The ITZ is a permeable region inside the multi-phase concrete, consisting of cement, aggregate, and the ITZ itself. Physio-chemical processes may decrease the high porosity of the ITZ. Enhancing the formation of a solid product of hydration (gel) and refilling the small openings would lead to the densification of the ITZ and a reduction in its pore volume. While dry rubberised concrete mixes do not have other types of holes such as those caused by purposeful air entrainment or inadequate compaction (CR concrete formed beneath a pressure of 2500 MPa), they do have a new kind of pores formed by the entrapment of air on the surface of the CR. Nevertheless, the gel has the ability to fill up these pores. Figure 4.2 demonstrates that the porosity of the rubberised concrete increases as the amount of CR substitution increases, which aligns with the findings of prior studies (Thomas and Gupta 2015). This occurs because of the lack of polarity in CR, which leads it to repel water and entangle air on its outermost layer. As a result, the air spaces in the matrix are improved, as well as the pores in the ITZ. Nevertheless, the incorporation of MWCNTs leads to a decrease in the porosity of CR concrete. MWCNTs serve as a filler to occupy empty spaces within the matrix and enhance the ITZ of rubberised concrete.

Furthermore, the overall number of pores in the rubberised concrete rises as the CR substitution rises, but reduces with the incorporation of MWCNTs, as seen in Fig. 4.2e and f. The corresponding rise in CR substitution leads to a rise in the porosity of rubberised concrete, as seen in Fig. 4.2b–d. However, the inclusion of MWCNTs mitigates this impact by lowering the porous volume of the rubberised concrete. Put simply, as the overall volume of holes in rubberised concrete grows, the porosity also rises, and vice versa. This suggests that the pores in the solidified rubberised concrete are linked, and the addition of MWCNTs aids in lowering the overall volume of holes and therefore decreasing the porosity. Figure 4.2a–f shows the distribution of pore sizes in rubberised concrete mixes, including varying amounts of MWCNTs (0%, 0.05%, 0.065%, and 0.08%) and variable levels of substitution of conventional rubber (CR) (0%, 1%, 3%, and 5%). The total pore volume rises as the CR concentration increased. The distribution of pore sizes ranges from 10 to 100,000 nm. Nevertheless, a greater pore volume within the range of pore sizes shows that the combination has a higher level of porosity and permeability. This is due to the fact that larger capillaries have a greater influence on permeability (Mohammed et al. 2016; Li et al. 2016; Du et al. 2014). The pore dimension spans from 10 to 100,000 nm in size. The results direct that as the proportion of MWCNTs rises, the total pore volume reduces. This

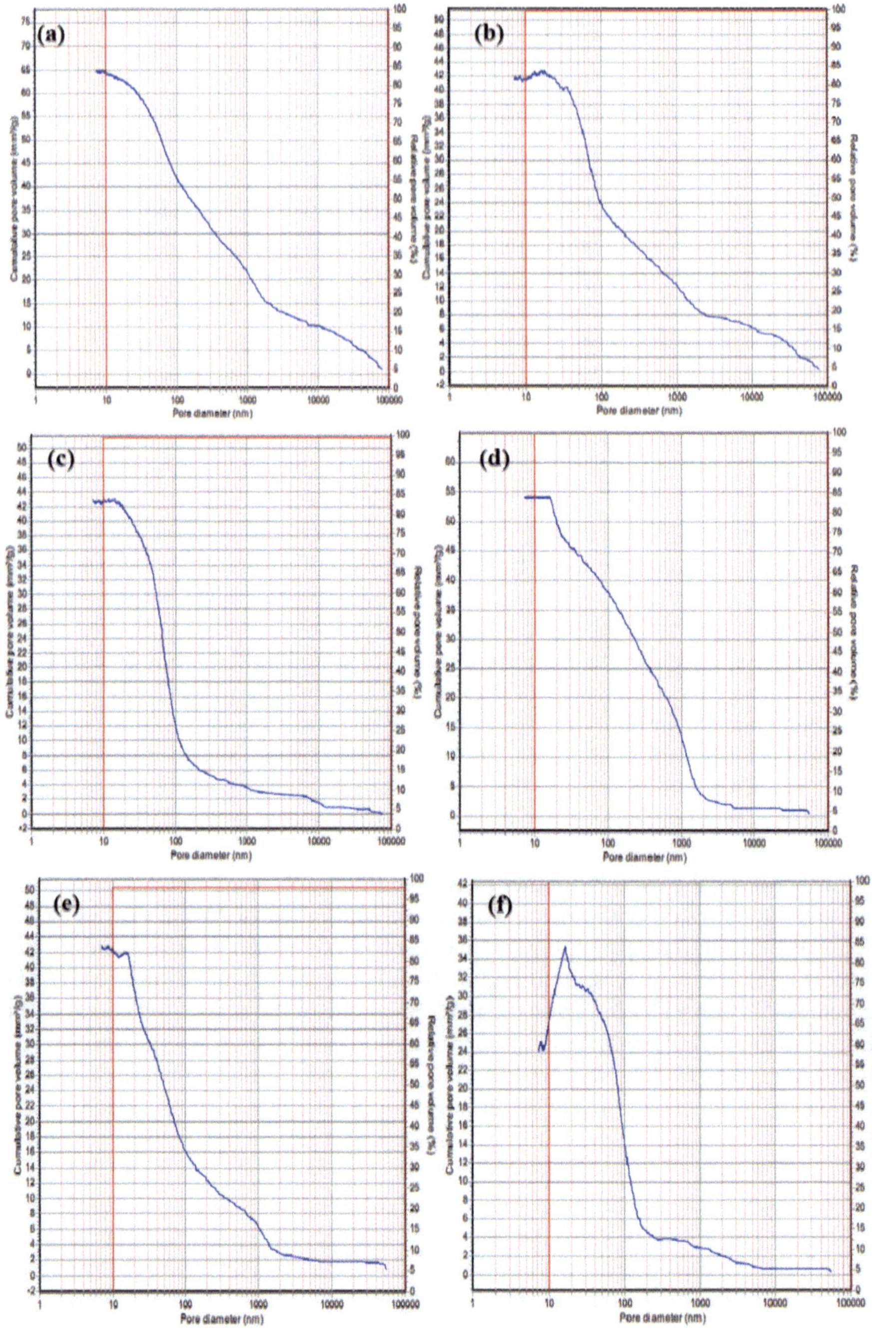

Fig. 4.2 MIP of rubberised concrete for **a** M0 (0%, MWCNTs, 0%, CR), **b** M9 (0.08% MWCNTs, 1% CR), **c** M6 (0.08% MWCNTs, 3% CR), **d** M2 (0.08% MWCNTs, 5% CR), **e** MCP (0.065% MWCNTs, 3% CR), and **f** M7 (0.05% MWCNTs, 1% CR)

may be attributed to the physio-chemical interactions of MWCNTs, which have the capacity to refine the pore structure, resulting in a reduction in the overall pore volume.

4.4 Response Surface Methodology (RSM) Modelling

4.4.1 ANOVA

The purpose of applying RSM is to develop models of response surfaces and evaluate them through ANOVA. The study focused on examining the CS of CR concrete with 0.05–0.08% MWCNTs at 7, 14, and 28 days for RSM modelling and optimisation purposes. Moreover, the study concluded that the 2F models were more appropriate for the CS. Furthermore, each of the replies given is expressed utilising encoded terminology, illustrated by Eqs. (4.1) to (4.3). Equations built using coded parts may be used to predict the outcome of various values for every parameter. The conventional notation for expressing the maximum values of parameters is displayed by + 1, while the minimum values are represented by − 1 (Khurshid et al. 2024; Mohammed et al. 2018d, 2021; Murali et al. 2021). The encoded mathematical models can be utilised to assess the comparative significance of the characteristics by analysing the factor coefficients. The study uses two input values, A and B, which represent MWCNTs and CR, consistently. Table 4.3 demonstrations the outcomes of the ANOVA analysis.

The ANOVA used a statistical significance threshold of 95%, through which any model constituent with a likelihood less than 5% was deemed statistically significant (Rahim et al. 2022; Bheel and Mohammed 2024; Bheel et al. 2023b). All the mathematical models developed in this context are statistically significant, as their likelihoods are less than 0.05. Each of the prototype correlations A, B, and AB, in the CS model has statistical significance. The R^2 is a critical metric for assessing the efficiency of an algorithm. The R^2 statistic assesses the appropriateness of the anticipated result given empirical study data, fluctuating from 0 to 1 or as a percentage from 0 to 100 (Appana et al. 2022; Adamu et al. 2017). Higher levels show more acceptability, whereas lower values indicate the opposite. Table 4.4 displays the R^2 value and other criteria used for model evaluation. The mathematical models' R^2 values were 97.19%, 97.42%, and 97.57% for CS at 7, 14, and 28 days, respectively, indicating a high level of accuracy. In addition, "Adeq. Precision" evaluates the ratio of the signal to the noise. A ratio more than 4 is preferable. The table reveals that the Adeq. Precision readings for the CS algorithms are 36.31, 37.74, and 39.03, at 7, 14, and 28 days, consistently. The results described above indicate that the mathematical methods used are adequate and have the capacity to appropriately predict responses.

$$\text{CS7} = +34.01 - 2.58 \times A - 2.98 \times B + 0.64 \times AB \tag{4.1}$$

Table 4.3 ANOVA findings of the CS

Response	Source	Sum of squares	*Df*	Mean square	*F*-value	*p*-Value > *F*	Significance
CS7	Model	94.79	3	31.60	103.72	< 0.0001	Yes
	A-MWCNTs	39.99	1	39.99	131.28	< 0.0001	Yes
	B-CR	53.16	1	53.16	174.52	< 0.0001	Yes
	AB	1.64	1	1.64	5.38	0.0456	Yes
	Residual	2.74	9	0.30			
	Lack of fit	2.74	5	0.55			
	Pure error	0.000	4	0.000			
	Cor total	97.53	12				
CS14	Model	124.96	3	41.65	113.28	< 0.0001	Yes
	A-MWCNTs	52.33	1	52.33	142.33	< 0.0001	Yes
	B-CR	69.09	1	69.09	187.90	< 0.0001	Yes
	AB	3.53	1	3.53	9.61	0.0127	Yes
	Residual	3.31	9	0.37			
	Lack of Fit	3.31	5	0.66			
	Pure error	0.000	4	0.000			
	Cor total	128.26	12				
CS28	Model	165.75	3	55.25	120.55	< 0.0001	Yes
	A-MWCNTs	71.55	1	71.55	156.13	< 0.0001	Yes
	B-CR	90.09	1	90.09	196.58	< 0.0001	Yes
	AB	4.10	1	4.10	8.95	0.0152	Yea
	Residual	4.12	9	0.46			
	Lack of fit	3.19	5	0.64	2.73	0.1763	No
	Pure error	0.94	4	0.23			
	Cor total	169.87	12				

$$\text{CS14} = +38.50 - 2.95 \times A - 3.39 \times B + 0.94 \times AB \tag{4.2}$$

$$\text{CS28} = +49.39 - 3.45 \times A - 3.88 \times B + 1.01 \times AB \tag{4.3}$$

Here, *A* indicates the MWCNTs, and *B* displays the CR, while CS7, CS14, and CS28 stand for CS at 7, 14, and 28 days, respectively.

The RSM assessment includes vital diagrams, such as important model diagnostic plots, which include the actual plot of observed versus expected values, which are utilised to verify the importance of the quadratic algorithms utilised for rubberised concrete containing MWCNTs. Figures 4.3, 4.4, and 4.5a show a significant correlation between the experimental and projected outcomes for CS models at 7, 14, and 28 days. The close proximity of spreading points and the small discrepancy

Table 4.4 Specifications for model validation

Model validation constraints	CS7	CS14	CS28
Mean	34.01	38.50	49.39
Std. Dev	0.55	0.61	0.68
C.V. %	1.62	1.57	1.37
PRESS	10.05	12.31	9.56
− 2 Log likelihood	16.66	19.11	21.97
Adj R^2	0.9625	0.9656	0.9676
R^2	0.9719	0.9742	0.9757
Predicted R^2	0.8969	0.9041	0.9437
Adeq precision	36.311	37.738	39.030

between investigational and projected values validate this concurrence. Significantly, the points exhibit a close alignment along a straight line, which suggests a normal distribution pattern and provides more evidence of the model's reliability. Besides, the figures labelled 4.3, 4.4, and 4.5b illustrate the relationships between input values and outcomes (responses) using 3D surface charts for 7, 14, and 28 days. They explicitly emphasise the connections among various input factors. The 3D response graphs for CS in Figs. 4.3, 4.4, and 4.5b show a notable concentration of CS at a 0.05% MWCNTs level compared to other levels of MWCNTs. Moreover, it is clear that the strength of rubberised concrete significantly increased when incorporating MWCNTs up to a concentration of 0.05%. This improvement may be accredited to the capability of MWCNTs, as nanoscale particles, to fill voids and increase the compactness of the CR mixture.

4.4.2 *Optimisation*

An optimisation effort is conducted to determine the optimal readings of each factor individually in order to achieve the highest level of the desired output. The technique entails setting objectives for many aspects, such as input parameters and output, while taking into account various requirements and levels of importance in order to accomplish the intended purpose. The assessment of optimisation is performed utilising the desirability rate, which ranges from 0 to 1 (Bheel et al. 2023c; d; Abdulkadir et al. 2021; Haruna et al. 2020; Udeze et al. 2024). A number closer to one indicates a more advantageous result.

Table 4.5 specifies the objectives and standards for the enhancement of this specific scenario. Table 4.5 shows that the optimisation method aims to maximise four distinct responses while also minimising one output. However, the use of MWCNTs has been restricted to a range of 0.05–0.08%, whereas the use of CR has been confined to a range of 1–5%. These limitations have been enforced to allow the system to figure out the optimal amount needed to accomplish the desired goal. The optimisation

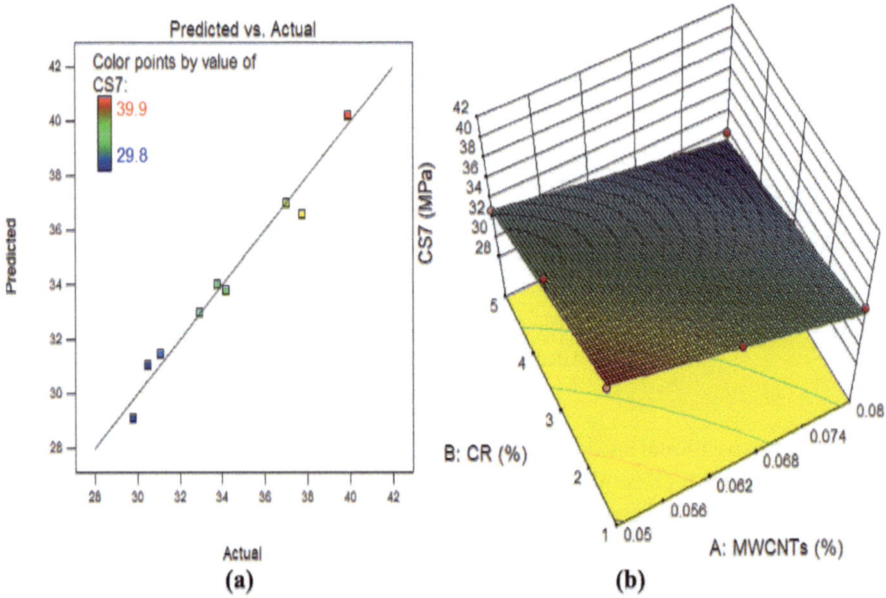

Fig. 4.3 **a** Actual against predicted plot and **b** 3D surface diagram for CS7

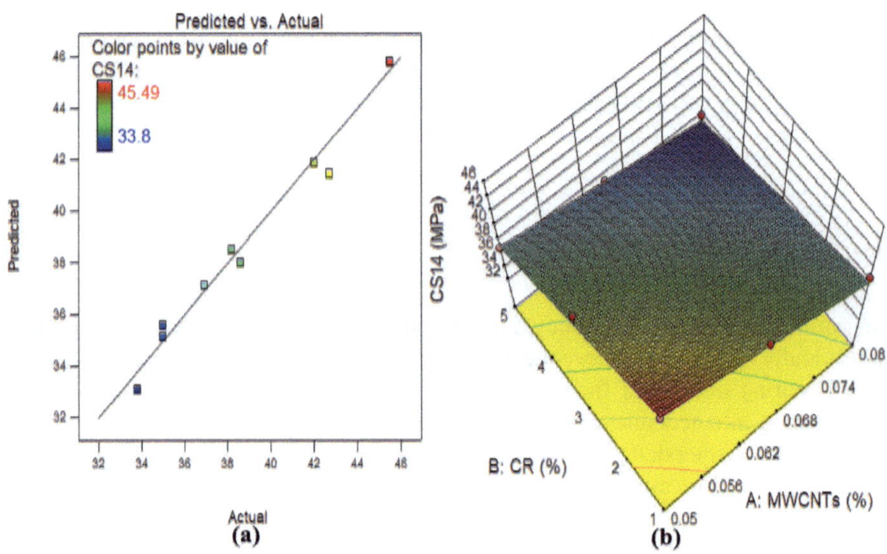

Fig. 4.4 **a** Actual against predicted plot and **b** 3D surface diagram for CS14

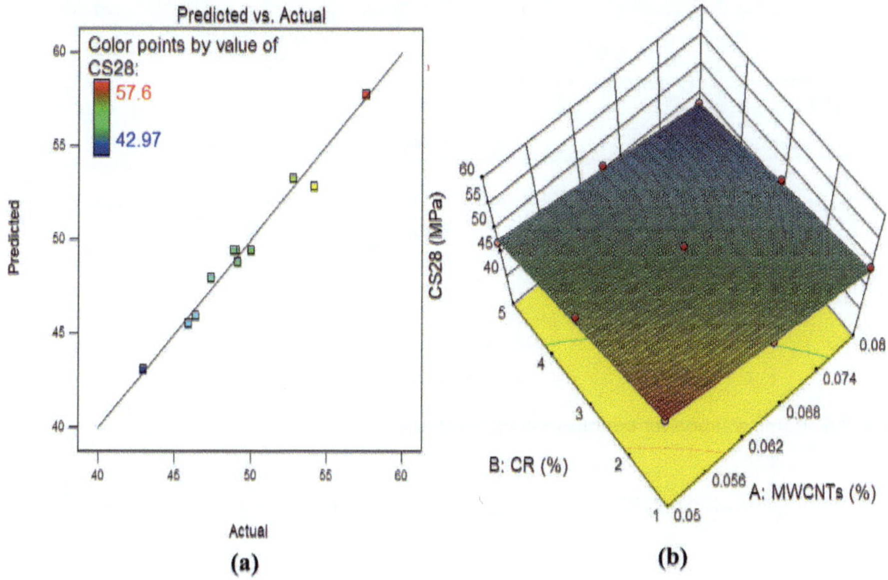

Fig. 4.5 **a** Actual against predicted plot and **b** 3D surface diagram for CS28

Table 4.5 Optimisation criteria of the outcomes

Factors		Input factors		Responses (compressive strength)		
		MWCNTs (%)	CR (%)	7 Days (MPa)	14 Days (MPa)	28 Days (MPa)
Value	Minimum	0.05	1	29.80	33.80	42.97
	Maximum	0.08	5	39.90	45.49	57.60
Goal		Range	Range	Maximise	Maximise	Maximise
Optimisation outcomes		0.05	1	40.12	45.67	57.60
Desirability		1.000 (100%)				

investigation revealed that adding 0.05% MWCNTs and 1% CR to the composites could increase the CS7, CS14, and CS28 readings to 40.12 MPa, 45.49 MPa, and 57.60 MPa, respectively. The optimisation is deemed to have a desirability of 100%, which is considered appropriate given the significant variability in the response values revealed in Fig. 4.6. Figure 4.7 depicts the individual 3D schematic for desirability.

4.4.3 Experimental Validation for Optimisation Outcomes

The optimised findings provide many potential solutions for achieving a desirability of approximately 1.00. Table 4.5 displays these solutions, categorising them

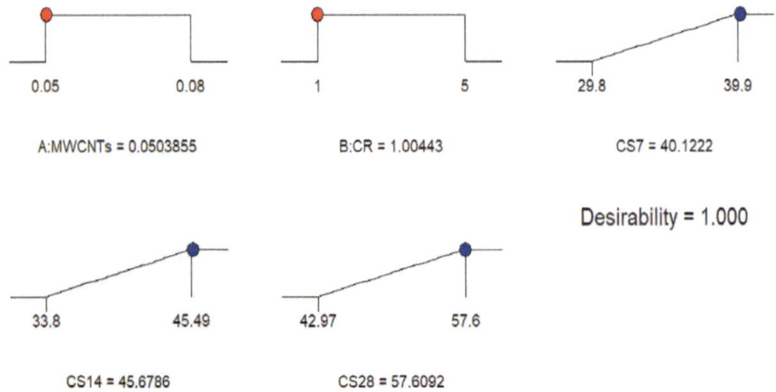

Fig. 4.6 Ramp solution for optimisation of outcomes

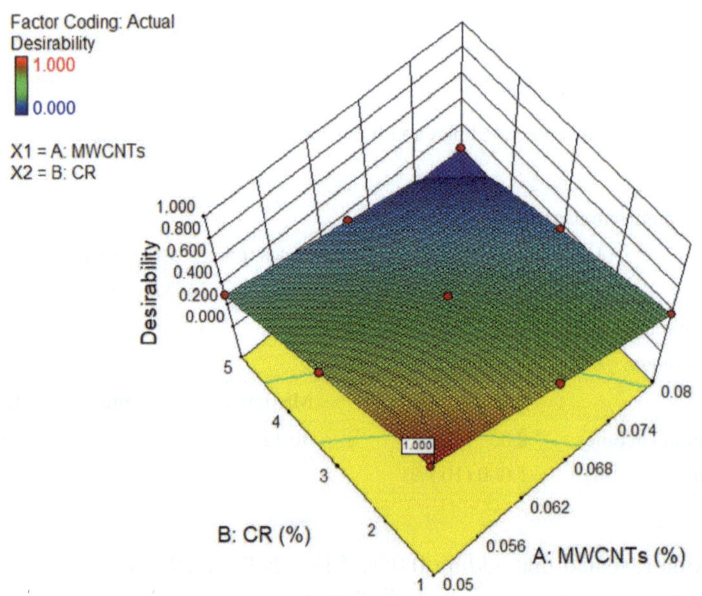

Fig. 4.7 3D Plot of desirability of optimisation outcomes

according to the optimum CS of rubberised concrete at 7, 14, and 28 days consistently. An experimental validation was performed for several of the optimised combinations mentioned in Table 4.6, yielding outcomes with a variance of lower than five per cent.

Table 4.6 Optimum outcomes of model validation

Responses	Practical outcomes	Predicted outcomes	Error (%)
CS 7 days (MPa)	39.70	40.12	1.05
CS 14 days (MPa)	45.20	45.67	1.03
CS 28 days (MPa)	57.40	57.60	0.35

4.5 Conclusions

This investigation aims to examine the effects of MWCNTs on rubberised concrete. The emphasis is planned on studying how MWCNTs affect the compressive strength and microstructural features of the concrete. Based on the acquired data, the resulting significant conclusions arise:

- The use of MWCNTs significantly enhanced the CS and microstructural characteristics of rubberised concrete. The increase in CS was particularly noticeable over the 28-day period. The most effective dose of MWCNTs, when used along with CR, was determined to be 0.05% and 1%, correspondingly.
- The improvements in the hardened properties of CR concrete resulted from the even distribution of MWCNTs. These nanoscale particles acted as places where cement hydrates could form, resulting in amended performance. Their large surface area accelerated the rate of hydration, thereby significantly contributing to the reported enhancements.
- The microstructural features of the rubberised concrete, when mixed with MWCNTs, were investigated by examination utilising MIP. The outcomes demonstrated that as the amount of MWCNTs rose, the microstructure of the concrete became more improved and finely structured. The presence of MWCNTs could potentially contribute to the reduction of pore size by filling the openings and promoting the production of crystalline hydration products.
- Response forecasting algorithms were developed and verified using both ANOVA and experimental validation. These algorithms have strong resilience, with coefficient of determination (R^2) levels ranging from 97 to 98%. Researchers determined that the most optimal values for the input variables are 0.05% for MWCNTs and 1% for CR using multi-objective optimisation. The optimisation procedure yielded a desirability value of 100%, indicating a satisfactory result.
- According to the research findings, using 0.05% MWCNTs in rubberised concrete produces the best outcomes, indicating that it is beneficial for use in the construction sector. This proposal is based on observable results that demonstrate its efficacy in real-world building situations.

References

Abdulkadir I, Mohammed BS, Liew MS, Wahab MMA (2021) Modelling and optimization of the impact resistance of graphene oxide modified crumb rubber-ECC using response surface methodology. IOP Conf Ser Mater Sci Eng 1197:012043. https://doi.org/10.1088/1757-899x/1197/1/012043

Abdulkadir I, Mohammed BS, Liew MS, Bin Abdul Wahab MM, Zawawi NAWA, As'ad S (2020) A review of the effect of waste tire rubber on the properties of ECC. Int J Adv Appl Sci 7:105–116. https://doi.org/10.21833/ijaas.2020.08.011

Abdulkadir I, Mohammed BS, Liew MS, Wahab MMA (2021) Modelling and optimization of the mechanical properties of engineered cementitious composite containing crumb rubber pretreated with graphene oxide using response surface methodology. Constr Build Mater 310. https://doi.org/10.1016/j.conbuildmat.2021.125259

Abdulkadir I, Mohammed BS, Liew MS, Wahab MMA (2021) Modelling and multi-objective optimization of the fresh and mechanical properties of self-compacting high volume fly ash ECC (HVFA-ECC) using response surface methodology (RSM). Case Stud Constr Mater 14. https://doi.org/10.1016/j.cscm.2021.e00525

Adamu M, Mohammed BS, Shafiq N (2017) Evaluating the effect of superplasticizer on the properties of roller compacted concrete using response surface methodology. ARPN J Eng Appl Sci 12:6206–6215

Adamu M, Mohammed BS, Shahir LM (2018) Mechanical properties and performance of high volume fly ash roller compacted concrete containing crumb rubber and nano silica. Constr Build Mater 171:521–538. https://doi.org/10.1016/j.conbuildmat.2018.03.138

Adamu M, Mohammed BS, Liew MS, Alaloul WS (2019) Evaluating the impact resistance of roller compacted concrete containing crumb rubber and nanosilica using response surface methodology and Weibull distribution. World J Eng 16:33–43. https://doi.org/10.1108/WJE-10-2018-0361

Adamu M, Mohammed BS, Shafiq N, Liew MS (2020) Durability performance of high volume fly ash roller compacted concrete pavement containing crumb rubber and nano silica. Int J Pavem Eng 21:1437–1444. https://doi.org/10.1080/10298436.2018.1547825

Adamu M, Mohammed BS, Shafiq N, Shahir Liew M (2018) Effect of crumb rubber and nano silica on the fatigue performance of roller compacted concrete pavement. Cogent Eng 5. https://doi.org/10.1080/23311916.2018.1436027

Adamu M, Mohammed BS, Liew MS (2018) Effect of crumb rubber and nano silica on the creep and drying shrinkage of roller compacted concrete pavement. Int J Geomate 15:58–65. https://doi.org/10.21660/2018.47.22082

Alaloul WS, Musarat MA, A Tayeh B, Sivalingam S, Rosli MF Bin, Haruna S et al (2020) Mechanical and deformation properties of rubberized engineered cementitious composite (ECC). Case Stud Constr Mater 13. https://doi.org/10.1016/j.cscm.2020.e00385

Albano C, Camacho N, Reyes J, Feliu JL, Hernández M (2005) Influence of scrap rubber addition to Portland I concrete composites: destructive and non-destructive testing. Compos Struct 71:439–446. https://doi.org/10.1016/j.compstruct.2005.09.037

Al-Fakih A, Mohammed B, Mohammed BS, Liew MS (2020) Tires rubber as a useable material in civil engineering applications: an overview. Int J Adv Res Eng Technol 11:315–325

Al-Fakih A, Mohammed BS, Liew MS (2021) On rubberized engineered cementitious composites (R-ECC): a review of the constituent material. Case Stud Constr Mater 14. https://doi.org/10.1016/j.cscm.2021.e00536

Appana PM, Mohammed BS, Abdulkadir I, Ali MOA, Liew MS (2022) Mechanical, microstructural and drying shrinkage properties of NaOH-pretreated crumb rubber concrete: RSM-based modeling and optimization. Materials 15. https://doi.org/10.3390/ma15072588

Assaggaf RA, Ali MR, Al-Dulaijan SU, Maslehuddin M (2021) Properties of concrete with untreated and treated crumb rubber—a review. J Mater Res Technol 11:1753–1798. https://doi.org/10.1016/j.jmrt.2021.02.019

ASTM C150 (2015) Standard specification for Portland cement. ASTM International, West Conshohocken, PA
Azmi NJ, Nasional UT (2008) Engineering properties of concrete containing recycled tire rubber. ICCBT 2008, pp 373–381
Bahari A, Berenjian J, Sadeghi-Nik A (2016) Modification of Portland cement with nano SiC. Proc Natl Acad Sci India Sect A Phys Sci 86:323–331. https://doi.org/10.1007/s40010-015-0244-y
Bheel N, Mohammed BS, Ali MOA, Shafiq N, Radu D (2023b) Effect of graphene oxide as a nanomaterial on the bond behaviour of engineered cementitious composites by applying RSM modelling and optimization. J Mater Res Technol 26:1484–1507. https://doi.org/10.1016/j.jmrt.2023.07.278
Bheel N, Mohammed BS (2024) Modelling and optimization of long-term modulus of elasticity and Poisson's ratio of graphene oxide based engineered cementitious composites by using response surface methodology. Diam Relat Mater 143. https://doi.org/10.1016/j.diamond.2024.110949
Bheel N, Mohammed BS, Liew MS, Zawawi NAWA (2023a) Durability behaviours of engineered cementitious composites blended with carbon nanotubes against sulphate and acid attacks by applying RSM modelling and optimization. Buildings 13. https://doi.org/10.3390/buildings13082032
Bheel N, Mohammed BS, Liew MS, Zawawi NAWA (2023c) Effect of graphene oxide as a nanomaterial on the durability behaviors of engineered cementitious composites by applying RSM modelling and optimization. Buildings 13. https://doi.org/10.3390/buildings13082026
Bheel N, Mohammed BS, Abdulkadir I, Liew MS, Zawawi NAWA (2023d) Effects of graphene oxide on the properties of engineered cementitious composites: multi-objective optimization technique using RSM. Buildings 13. https://doi.org/10.3390/buildings13082018
Bheel N, Mohammed BS, Mohamad H, Sutanto MH, Tafsirojjaman T (2024) Synergetic effect of multiwalled carbon nanotubes on mechanical and deformation properties of engineered cementitious composites: RSM modelling and optimization. Diam Relat Mater 147. https://doi.org/10.1016/j.diamond.2024.111299
Bravo M, De Brito J (2012) Concrete made with used tyre aggregate: durability-related performance. J Clean Prod 25:42–50. https://doi.org/10.1016/j.jclepro.2011.11.066
BS EN 12390-3 (2009) Testing hardened concrete. In: Compressive strength of test specimens. Standards Institution, London
Carriço A, Bogas JA, Hawreen A, Guedes M (2018) Durability of multi-walled carbon nanotube reinforced concrete. Constr Build Mater 164:121–133. https://doi.org/10.1016/j.conbuildmat.2017.12.221
Chaturvedy GK, Pandey UK (2022) Performance characteristics of rubberized concrete: a multipoint review. Innov Infrastruct Solut 7. https://doi.org/10.1007/s41062-021-00637-3
Chaturvedy GK, Pandey UK, Singh MP (2023) Analyzing the behavior of graphene oxide on high-strength rubberized concrete properties using different optimization techniques. Diam Relat Mater 140. https://doi.org/10.1016/j.diamond.2023.110485
Chou LH, Yang CK, Lee MT, Shu CC (2010) Effects of partial oxidation of crumb rubber on properties of rubberized mortar. Compos Part B Eng 41:613–616. https://doi.org/10.1016/j.compositesb.2010.09.009
Demir F, Yesilata B, Turgut P, Bulut H, Isiker Y (2015) Investigation of the effects of pH, aging and scrap tire content on the dissolution behaviors of new scrap tire-concrete mixture structures. J Clean Prod 93:38–46. https://doi.org/10.1016/j.jclepro.2015.01.043
Dong Q, Huang B, Shu X (2013) Rubber modified concrete improved by chemically active coating and silane coupling agent. Constr Build Mater 48:116–123. https://doi.org/10.1016/j.conbuildmat.2013.06.072
Du H, Du S, Liu X (2014) Durability performances of concrete with nano-silica. Constr Build Mater 73:705–712. https://doi.org/10.1016/j.conbuildmat.2014.10.014
Eftekhari M, Mohammadi S (2016) Multiscale dynamic fracture behavior of the carbon nanotube reinforced concrete under impact loading. Int J Impact Eng 87:55–64. https://doi.org/10.1016/j.ijimpeng.2015.06.023

Eltayeb E, Ma X, Zhuge Y, Xiao J, Youssf O (2021) Dynamic performance of rubberised concrete and its structural applications—an overview. Eng Struct 234. https://doi.org/10.1016/j.engstruct.2021.111990

Faried AS, Mostafa SA, Tayeh BA, Tawfik TA (2021) Mechanical and durability properties of ultra-high performance concrete incorporated with various nano waste materials under different curing conditions. J Build Eng 43. https://doi.org/10.1016/j.jobe.2021.102569

Fauzan NOF, Albarqi K, Melinda AP, Jauhari ZA (2021) The effect of waste tyre rubber on mechanical properties of normal concrete and fly ash concrete. Int J Geomate 20:55–61. https://doi.org/10.21660/2020.77.5737

Ganjian E, Khorami M, Maghsoudi AA (2009) Scrap-tyre-rubber replacement for aggregate and filler in concrete. Constr Build Mater 23:1828–1836. https://doi.org/10.1016/j.conbuildmat.2008.09.020

Gong K, Asce SM, Pan Z, Korayem AH, Ph D, Qiu L et al (2014) Reinforcing effects of graphene oxide on Portland cement paste. J Mater Civ Eng 27:1–6

Haruna S, Mohammed BS, Wahab MMA, Haruna A (2020) Compressive strength and workability of high calcium one-part alkali activated mortars using response surface methodology. IOP Conf Ser Earth Environ Sci 476. https://doi.org/10.1088/1755-1315/476/1/012018

Hong DLH, Mohammed BS, Al-Fakih A, Wahab MMA, Liew MS, Mugahed Amran YH (2020) Deformation properties of rubberized ECC incorporating nano graphene using response surface methodology. Materials 13:1–14. https://doi.org/10.3390/ma13122831

Huang B, Shu X, Cao J (2013) A two-staged surface treatment to improve properties of rubber modified cement composites. Constr Build Mater 40:270–274. https://doi.org/10.1016/j.conbuildmat.2012.11.014

Iijishima S (1991) Helical microtubules of graphitic carbon. Nature 354:56–58

Khan S, Ashraf S, Ali S, Khan K (2023) Experimental investigations on the mechanical properties and damage detection of carbon nanotubes modified crumb rubber concrete. J Build Eng 75. https://doi.org/10.1016/j.jobe.2023.106937

Khed VC, Mohammed BS, Liew MS, Abdullah Zawawi NAW (2020) Development of response surface models for self-compacting hybrid fibre reinforced rubberized cementitious composite. Constr Build Mater 232. https://doi.org/10.1016/j.conbuildmat.2019.117191

Khurshid H, Mohammed BS, Bheel N, Cahyadi WA, Mukhtar H (2024) Study of factors affecting the magnetic sensing capability of shape memory alloys for non-destructive evaluation of cracks in concrete: using response surface methodology (RSM) and artificial neural network (ANN) approaches. Heliyon 10. https://doi.org/10.1016/j.heliyon.2024.e35772

Konsta-Gdoutos MS, Metaxa ZS, Shah SP (2010) Multi-scale mechanical and fracture characteristics and early-age strain capacity of high performance carbon nanotube/cement nanocomposites. Cem Concr Compos 32:110–115. https://doi.org/10.1016/j.cemconcomp.2009.10.007

Kundan P, Sharma S (2020) Rubberized cemented concrete composites: a review. Mater Today Proc 44:4838–4842. https://doi.org/10.1016/j.matpr.2020.11.696

Lee J, Mahendra S, Alvarez PJJ (2010) Nanomaterials in the construction industry: a review of their applications and environmental health and safety considerations. ACS Nano 4:3580–3590. https://doi.org/10.1021/nn100866w

Li G, Wang Z, Leung CKY, Tang S, Pan J, Huang W et al (2016) Properties of rubberized concrete modified by using silane coupling agent and carboxylated SBR. J Clean Prod 112:797–807. https://doi.org/10.1016/j.jclepro.2015.06.099

Li Y, Li H, Wang Z, Jin C (2020) Effect and mechanism analysis of functionalized multi-walled carbon nanotubes (MWCNTs) on C-S-H gel. Cem Concr Res 128. https://doi.org/10.1016/j.cemconres.2019.105955

Li Y, Zhang X, Wang R, Lei Y (2019) Performance enhancement of rubberised concrete via surface modification of rubber: a review. Constr Build Mater 227. https://doi.org/10.1016/j.conbuildmat.2019.116691

Liew KM, Kai MF, Zhang LW (2016) Carbon nanotube reinforced cementitious composites: an overview. Compos Part A Appl Sci Manuf 91:301–323. https://doi.org/10.1016/j.compositesa.2016.10.020

Lv S, Ma Y, Qiu C, Sun T, Liu J, Zhou Q (2013) Effect of graphene oxide nanosheets of microstructure and mechanical properties of cement composites. Constr Build Mater 49:121–127. https://doi.org/10.1016/j.conbuildmat.2013.08.022

Meddah A, Beddar M, Bali A (2014) Use of shredded rubber tire aggregates for roller compacted concrete pavement. J Clean Prod 72:187–192. https://doi.org/10.1016/j.jclepro.2014.02.052

Mohammed BS (2010) Structural behavior and m-k value of composite slab utilizing concrete containing crumb rubber. Constr Build Mater 24:1214–1221. https://doi.org/10.1016/j.conbuildmat.2009.12.018

Mohammed BS, Adamu M (2018) Mechanical performance of roller compacted concrete pavement containing crumb rubber and nano silica. Constr Build Mater 159:234–251. https://doi.org/10.1016/j.conbuildmat.2017.10.098

Mohammed BS, Azmi NJ, Abdullahi M (2011) Evaluation of rubbercrete based on ultrasonic pulse velocity and rebound hammer tests. Constr Build Mater 25:1388–1397. https://doi.org/10.1016/j.conbuildmat.2010.09.004

Mohammed BS, Anwar Hossain KM, Eng Swee JT, Wong G, Abdullahi M (2012) Properties of crumb rubber hollow concrete block. J Clean Prod 23:57–67. https://doi.org/10.1016/j.jclepro.2011.10.035

Mohammed A, Sanjayan JG, Duan WH, Nazari A (2015) Incorporating graphene oxide in cement composites: a study of transport properties. Constr Build Mater 84:341–347. https://doi.org/10.1016/j.conbuildmat.2015.01.083

Mohammed BS, Awang AB, Wong SS, Nhavene CP (2016) Properties of nano silica modified rubbercrete. J Clean Prod 119:66–75. https://doi.org/10.1016/j.jclepro.2016.02.007

Mohammed BS, Adamu M, Shafiq N (2017) A review on the effect of crumb rubber on the properties of rubbercrete. Int J Civ Eng Technol 8:599–615

Mohammed BS, Khed VC, Nuruddin MF (2018a) Rubbercrete mixture optimization using response surface methodology. J Clean Prod 171:1605–1621. https://doi.org/10.1016/j.jclepro.2017.10.102

Mohammed BS, Adamu M, Liew MS (2018b) Evaluating the effect of crumb rubber and nano silica on the properties of high volume fly ash roller compacted concrete pavement using non-destructive techniques. Case Stud Constr Mater 8:380–391. https://doi.org/10.1016/j.cscm.2018.03.004

Mohammed BS, Liew MS, S Alaloul W, Al-Fakih A, Ibrahim W, Adamu M (2018c) Development of rubberized geopolymer interlocking bricks. Case Stud Constr Mater 8:401–408. https://doi.org/10.1016/j.cscm.2018.03.007

Mohammed BS, Adamu M, Liew MS (2018) Evaluating the static and dynamic modulus of elasticity of roller compacted rubbercrete using response surface methodology. Int J Geomate 14:186–192. https://doi.org/10.21660/2018.41.42833

Mohammed BS, Xian LW, Haruna S, Liew MS, Abdulkadir I, Zawawi NAWA (2021) Deformation properties of rubberized engineered cementitious composites using response surface methodology. Iran J Sci Technol Trans Civ Eng 45:729–740. https://doi.org/10.1007/s40996-020-00444-3

Mohammed BS, Yen LY, Haruna S, Seng Huat ML, Abdulkadir I, Al-Fakih A et al (2020) Effect of elevated temperature on the compressive strength and durability properties of crumb rubber engineered cementitious composite. Materials 13. https://doi.org/10.3390/MA13163516

Mousavi MA, Sadeghi-Nik A, Bahari A, Jin C, Ahmed R, Ozbakkaloglu T et al (2021) Strength optimization of cementitious composites reinforced by carbon nanotubes and Titania nanoparticles. Constr Build Mater 303. https://doi.org/10.1016/j.conbuildmat.2021.124510

Murali M, Mohammed BS, Abdulkadir I, Liew MS, Alaloul WS (2021) Utilization of crumb rubber and high-volume fly ash in concrete for environmental sustainability: RSM-based modeling and optimization. Materials 14. https://doi.org/10.3390/ma14123322

Nam IW, Park SM, Lee HK, Zheng L (2017) Mechanical properties and piezoresistive sensing capabilities of FRP composites incorporating CNT fibers. Compos Struct 178:1–8. https://doi.org/10.1016/j.compstruct.2017.07.008

Nazari A, Riahi S, Riahi S, Shamekhi SF, Khademno A (2010) Influence of Al_2O_3 nanoparticles on the compressive strength and workability of blended concrete. J Am Sci 6:6–9

Neville AM (2011) Properties of concrete. Pearson Education Limited, England

Nik AS, Bahari A, Ebadi AG, Nik AS, Ghasemi-Hamzekolaee A (2010) The role of nano particles (Si) in gate dielectric. Indian J Sci Technol 3:634–636. https://doi.org/10.17485/ijst/2010/v3i6/29771

Onuaguluchi O, Panesar DK (2014) Hardened properties of concrete mixtures containing pre-coated crumb rubber and silica fume. J Clean Prod 82:125–131. https://doi.org/10.1016/j.jclepro.2014.06.068

Pan Z, He L, Qiu L, Korayem AH, Li G, Zhu JW et al (2015) Mechanical properties and microstructure of a graphene oxide-cement composite. Cem Concr Compos 58:140–147. https://doi.org/10.1016/j.cemconcomp.2015.02.001

Pelisser F, Zavarise N, Longo TA, Bernardin AM (2011) Concrete made with recycled tire rubber: effect of alkaline activation and silica fume addition. J Clean Prod 19:757–763. https://doi.org/10.1016/j.jclepro.2010.11.014

Rahim NI, Mohammed BS, Abdulkadir I, Dahim M (2022) Effect of crumb rubber, fly ash, and nanosilica on the properties of self-compacting concrete using response surface methodology. Materials 15. https://doi.org/10.3390/ma15041501

Rostami H, Lepore J, Silverstraim T, Zundi I (2000) Use of recycled rubber tires in concrete. Proc Int Conf Concr 391–399

Sabapathy L, Mohammed BS, Al-Fakih A, Wahab MMA, Liew MS, Amran YHM (2020) Acid and sulphate attacks on a rubberized engineered cementitious composite containing graphene oxide. Materials 13. https://doi.org/10.3390/ma13143125

Sadek DM, El-Attar MM (2015) Structural behavior of rubberized masonry walls. J Clean Prod 89:174–186. https://doi.org/10.1016/j.jclepro.2014.10.098

Segre N, Joekes I (2000) Use of tire rubber particles as addition to cement paste. Cem Concr Res 30:1421–1425. https://doi.org/10.1016/S0008-8846(00)00373-2

Shu X, Huang B (2014) Recycling of waste tire rubber in asphalt and portland cement concrete: an overview. Constr Build Mater 67:217–224. https://doi.org/10.1016/j.conbuildmat.2013.11.027

Sldozian RJ, Mikhaleva Z, Tkachev A (2019) Evaluation of the efficiency of lightweight concrete modified with additives based on nanostructures. IOP Conf Ser Mater Sci Eng 693. https://doi.org/10.1088/1757-899X/693/1/012009

Thomas BS, Gupta RC (2015) Long term behaviour of cement concrete containing discarded tire rubber. J Clean Prod 102:78–87. https://doi.org/10.1016/j.jclepro.2015.04.072

Thomas BS, Gupta RC, Panicker VJ (2016) Recycling of waste tire rubber as aggregate in concrete: durability-related performance. J Clean Prod 112:504–513. https://doi.org/10.1016/j.jclepro.2015.08.046

Udeze OJ, Mohammed BS, Adebanjo AU, Abdulkadir I (2024) Optimizing an eco-friendly high-density concrete for offshore applications: a study on fly ash partial replacement and graphene oxide nano reinforcement. Case Stud Chem Environ Eng 9. https://doi.org/10.1016/j.cscee.2023.100592

Wang J, Dong S, Pang SD, Zhou C, Han B (2022) Pore structure characteristics of concrete composites with surface-modified carbon nanotubes. Cem Concr Compos 128. https://doi.org/10.1016/j.cemconcomp.2022.104453

Youssf O, Elgawady MA, Mills JE, Ma X (2014) An experimental investigation of crumb rubber concrete confined by fibre reinforced polymer tubes. Constr Build Mater 53:522–532. https://doi.org/10.1016/j.conbuildmat.2013.12.007

Zhang Z, Ma H, Qian S (2015) Investigation on properties of ECC incorporating crumb rubber of different sizes. J Adv Concr Technol 13:241–251. https://doi.org/10.3151/jact.13.241

Zhang P, Su J, Guo J, Hu S (2023) Influence of carbon nanotube on properties of concrete: a review. Constr Build Mater 369. https://doi.org/10.1016/j.conbuildmat.2023.130388
Zheng M, Liu Y, Liu X, Zhang W, Wang F, Liu S (2021) Study on the viscoelastic behaviour of the modified asphalt containing multi-walled carbon nanotubes (MWCNTs) and crumb rubber (CR). Constr Build Mater 311. https://doi.org/10.1016/j.conbuildmat.2021.125244

Chapter 5
Complex Fermatean Fuzzy Graph and Its Application in Networking Problems

Broumi Said, S. Mohanaselvi, Assia Bakali, and Mohamed Talea

Abstract Complex fuzzy sets (CFSs) are efficient tool to handle the uncertainties present in real-world complex-valued physical quantities and have become a prominent research topic in fuzzy set theory. The CFS is used to handle information about uncertainty and periodicity simultaneously in decision-making problem under complex environment. The aim of this research is to establish the innovative concept of complex Fermatean fuzzy graph (CFFG) and its scope of application in network optimization problems. CFFG is a generalization of a complex Pythagorean fuzzy graph (CPyFG) that is used to handle the vagueness with the degrees whose ranges are enlarged from real to complex subset with unit disc. In this paper, we studied a complex Fermatean fuzzy shortest path problem (CFFSPP) where the edge weights of the graph take complex Fermatean fuzzy numbers (CFFN) instead of real numbers. We developed a generalized Dijkstra's algorithm using the existing arithmetic operation and improved score function of CFFN to find the shortest path. A numerical example is discussed to showcase the potency of the proposed algorithm.

Keywords Complex fuzzy set · Fermatean fuzzy set · Improved score function · SPP

B. Said (✉) · M. Talea
Laboratory of Information Processing, Faculty of Science Ben M'Sik, University of Hassan II, Casablanca, Morocco
e-mail: broumisaid78@gmail.com

M. Talea
e-mail: Taleamohamed@yahoo.fr

B. Said
STIE Team Regional Center for the Professions of Education and Training (C.R.M.E.F), Casablanca, Settat, Morocco

S. Mohanaselvi
Department of Mathematics, SRM Institute of Science and Technology, Kattankulathur, India
e-mail: mohanass@srmist.edu.in

A. Bakali
Ecole Royale Navale, Casablanca, Morocco
e-mail: assiabakali@yahoo.fr

B. Shukla et al. (eds.), *Sustainable Technologies for Transforming Businesses and Societies*, Transactions on Computer Systems and Networks,
https://doi.org/10.1007/978-981-96-9471-6_5

5.1 Introduction

The real-world uncertainty in decision-making process is very well handled by fuzzy sets (FSs) introduced by Zadeh (1965) as it replicates the human reasoning. But FS only considers the membership degree (α) and does not allow us to select independently non-membership degree (β). Atanassov (1986) proposed Intuitionistic fuzzy sets [IFS] which facilitates the decision-maker (DM) to assign α and β independently with the restriction $0 \le \alpha + \beta \le 1$. Yager and Abbasov (2013) and Yager (2014) introduced the Pythagorean fuzzy sets [PFS} by improving the restriction of IFS to $0 \le \alpha^2 + \beta^2 \le 1$. As IFS and PFS restrict the DM to select α and β in a certain range, Sanapati et al. (2019) and Senapati and Yager (2019) proposed Fermatean fuzzy set (FFS) with the restriction $0 \le \alpha^3 + \beta^3 \le 1$ to handle a wide range of uncertainty.

Many physical quantities in practical problems, such as wave function in quantum mechanics, impedance in electrical engineering, etc., are complex-valued. Buckley (1989) first initiated the formal definition of complex fuzzy numbers (CFN) and its types. Later on, in 2002 Ramot et al. (2002) proposed the innovative concept of complex fuzzy sets by extending fuzzy sets (FSs) in real field to complex field. That is for CFS the membership function's range [0,1] in traditional FS is extended to the unit circle in the complex plane. The membership function of CFS is a complex-valued function whose membership degree is of the form $\alpha e^{i\omega}$ where $= \sqrt{-1}$. CFS allows us to describe the membership degree in two terms, namely amplitude term α which lies in [0,1] and phase term ω which lies in the interval [0,2π]. Many researchers have focused their attention on CFS over the past few years. Yazdanbakhsh and Dick (2018) made a systematic review of the development of CFS and highlighted possible future directions for research. Many researchers investigated and worked on CFS and its extensions. Alkouri and Salleh (2012) generalized the concept of CFS to IFS and proposed complex intuitionistic fuzzy set (CIFS). CIFS is characterized by a complex-valued membership and non-membership function. Greenfield et al. (2016) brought up the innovative idea of complex interval valued fuzzy set (CInVFS). Ullah et al. (2019) developed the idea of CPyFS. Akram et al. (2020) proposed a new concept called as complex picture fuzzy set (CPFS) and includes applications in multi criteria decision-making (MCDM) problems. Chinnadurai et al. (2021) presented complex Fermatean fuzzy set (CFFS) in rectangular coordinates and its applications in MCDM problems. Garg et al. (2021) proposed the concept of complex hesitant fuzzy set and its measures. Extensions of CFS are shown in Table 5.1.

Fuzzy graphs (FG) are developed to model uncertainties in graphical network, and its extensions are explored by many researchers (Luqman et al. 2019; Shannon and Atanassov 1994; Thamizhendhi et al. 2011). Thirunavukarasu et al. (2016) extended FG to complex fuzzy graphs (CFG) and proposed energy of CFG. Yaqoob and Akram (2018) studied about complex neutrosophic graphs (CNG). Also, Yaqoob et al. (2019) proposed the new idea of complex intuitionistic fuzzy graphs (CIFG) using some basic operations. Akram and Naz (2019) proposed complex Pythagorean fuzzy graphs (CPFGs) and their properties. Zeng et al. (2020) proposed the complex vague graph (CVG) as a generalization of CFG. Recently, Shoaib et al. (2022) proposed

Table 5.1 Complex fuzzy sets under various fuzzy environment

References	Year	Significance influences
Ramot et al. (2002)	2002	CFS
Alkouri and Salleh (2012)	2012	CIFS
Greenfield et al. (2016)	2016	CInVFS
Ullah et al. (2019)	2019	CPyFS
Akram et al. (2020)	2020	CPFS
Chinnadurai et al. (2021)	2021	CFFS
Garg et al. (2021)	2021	CHFS

the concept of complex spherical fuzzy graph and its operations. The extensions of CFG are shown in Table 5.2.

Graph models have many applications in networking problems. A thorough review of the prior SPP literature (Gani and Jabarulla 2010; Baba 2013; Khorsandi et al. 2016; Jan et al. 2019) reveals that IFS and PFS environment are the only topics that are discussed. This study's goal is to implement SPP in a complex Fermatean fuzzy (CFF) environment.

The main contributions of this article are as follows:

1. First and foremost, we initiated the new idea of complex Fermatean fuzzy graph.
2. We explored shortest path problem under complex environment and modelled CFFSPP.
3. A generalized Dijkstra's algorithm for solving CFFSPP is proposed based on the arithmetic operations and score function of CFFNs.
4. The feasibility and viability of proposed algorithm are discussed using a numerical example.

The rest of this paper is organized as follows. In Sect. 5.2, we review preliminaries with respect to FFS. In Sect. 5.3, a new algorithm based on traditional Dijkstra's algorithm for CFFSPP is developed to obtain the shortest path. In Sect. 5.4, an illustrative example is discussed. Finally, conclusions and the future work of the proposed algorithm are discussed in Sect. 5.5.

Table 5.2 CFG under various fuzzy environment

References	Year	Significance influences
Thirunavukarasu et al. (2016)	2016	CFG
Yaqoob and Akram (2018)	2018	CNG
Yaqoob et al. (2019)	2019	CIFG
Akram and Naz (2019)	2019	CPyFG
Zeng et al. (2020)	2020	CVG
Shoaib et al. (2022)	2022	CSFG
Proposed		CFFG

5.2 Preliminaries

Let us give a brief review of some essential concepts that are concerned with the present study.

Definition 2.1 (Alkouri and Salleh 2012)

A CIFS $\hat{\mathbb{I}}$ defined on a universal set $\mathbb{X}$ is a structure having the form as

$$\hat{\mathbb{I}} = \{\mathrm{a}, \alpha_{\hat{\mathbb{I}}}(\mathrm{a}) = \mu, \gamma_{\hat{\mathbb{I}}}(\mathrm{a}) = \mu' : \mathrm{a} \in \mathbb{X}\}$$

where $\alpha_{\hat{\mathbb{I}}} : \mathbb{X} \to \{\mu | \mu \in \mathbb{C}, |\mu| \leq 1\}$ indicates the degree of membership and $\beta_{\hat{\mathbb{I}}} : \mathbb{X} \to \{\mu' | \mu' \in \mathbb{C}, |\mu'| \leq 1\}$ indicates the degree of non-membership of every element $a\varepsilon\mathbb{X}$ such that $\alpha_{\hat{\mathbb{I}}}(a) = \mu = x_1 + iy_1$ and $\beta_{\hat{\mathbb{I}}}(a) = \mu' = x_2 + iy_2$ provided $0 \leq |\mu| + |\mu'| \leq 1$ or $\alpha_{\hat{\mathbb{I}}}(a) = r_{\hat{\mathbb{I}}}(a) \cdot e^{i\omega_{\hat{\mathbb{I}}}(a)}$ and $\beta_{\hat{\mathbb{I}}}(a) = k_{\hat{\mathbb{I}}}(a) \cdot e^{i\rho_{\hat{\mathbb{I}}}(a)}$such that $0 \leq r_{\hat{\mathbb{I}}}(a), k_{\hat{\mathbb{I}}}(a) \leq 1; 0 \leq \omega_{\hat{\mathbb{I}}}(a), \rho_{\hat{\mathbb{I}}}(a) \leq 2\pi$ satisfying the condition $0 \leq r_{\hat{\mathbb{I}}}(a) + k_{\hat{\mathbb{I}}}(a) \leq 1$, also $\omega_{\hat{\mathbb{I}}}$ and $\rho_{\hat{\mathbb{I}}}$ are real − valued. That is CIFS $\hat{\mathbb{I}}$ assigns any element $a\varepsilon\mathbb{X}$ a complex-valued grade of both membership and non-membership. Furthermore, a complex intuitionistic fuzzy number (CIFN) is given by. $\delta = \left(r \cdot e^{2\pi i\omega_{\hat{\mathbb{I}}}}, k \cdot e^{2\pi i\rho_{\hat{\mathbb{I}}}}\right)$.

Definition 2.2 (Ullah et al. 2019)

A CPyFS $\hat{\mathbb{P}}$ defined on a universal set $\mathbb{X}$ is a structure having the form as

$$\hat{\mathbb{P}} = \{\mathrm{a}, \mathrm{r}_{\hat{\mathbb{P}}}(\mathrm{a}) \cdot \mathrm{e}^{\mathrm{i}\omega_{\hat{\mathbb{P}}}(\mathrm{a})}, \mathrm{k}_{\hat{\mathbb{P}}}(\mathrm{a}) \cdot \mathrm{e}^{\mathrm{i}\rho_{\hat{\mathbb{P}}}(\mathrm{a})} : \mathrm{a} \in \mathbb{X}\}$$

where $r_{\hat{\mathbb{P}}}(a), k_{\hat{\mathbb{P}}}(a) \in [0, 1], 0 \leq r^2_{\hat{\mathbb{P}}}(a) + k^2_{\hat{\mathbb{P}}}(a) \leq 1; \omega_{\hat{\mathbb{P}}}(a), \rho_{\hat{\mathbb{P}}}(a) \in [0, 2\pi]$ with $0 \leq \frac{\omega^2_{\hat{\mathbb{P}}}(a)}{2\pi} + \frac{\rho^2_{\hat{\mathbb{P}}}(a)}{2\pi} \leq 1$. The amplitude terms of CPyFS are $r_{\hat{\mathbb{P}}}, k_{\hat{\mathbb{P}}}$ and the phase terms are $\omega_{\hat{\mathbb{P}}}, \rho_{\hat{\mathbb{P}}}$. Furthermore, a complex intuitionistic fuzzy number (CIFN) is given by $\delta = \left(r \cdot e^{2\pi i\omega_{\hat{\mathbb{P}}}}, k \cdot e^{2\pi i\rho_{\hat{\mathbb{P}}}}\right)$.

Definition 2.3 (Chinnadurai et al. 2021).

A CFFS $\hat{\mathbb{F}}$ defined on a universal set $\mathbb{X}$ is a structure having the form as

$$\hat{\mathbb{F}} = \{\mathrm{a}, \mathrm{r}_{\hat{\mathbb{F}}}(\mathrm{a}) \cdot \mathrm{e}^{\mathrm{i}\omega_{\hat{\mathbb{F}}}(\mathrm{a})}, \mathrm{k}_{\hat{\mathbb{P}}}(\mathrm{a}) \cdot \mathrm{e}^{\mathrm{i}\rho_{\hat{\mathbb{F}}}(\mathrm{a})} : \mathrm{a} \in \mathbb{X}\}$$

where $r_{\hat{\mathbb{F}}}(a), k_{\hat{\mathbb{F}}}(a) \in [0, 1], 0 \leq r^3_{\hat{\mathbb{F}}}(a) + k^3_{\hat{\mathbb{F}}}(a) \leq 1; \omega_{\hat{\mathbb{F}}}(a), \rho_{\hat{\mathbb{F}}}(a) \in [0, 2\pi]$ With $0 \leq \frac{\omega^3_{\hat{\mathbb{F}}}(a)}{2\pi} + \frac{\rho^3_{\hat{\mathbb{F}}}(a)}{2\pi} \leq 1$. The amplitude terms of CFFS are $r_{\hat{\mathbb{F}}}, k_{\hat{\mathbb{F}}}$ and the phase terms are $\omega_{\hat{\mathbb{F}}}, \rho_{\hat{\mathbb{F}}}$. Furthermore, a complex intuitionistic fuzzy number (CIFN) is given by $\delta = \left(r \cdot e^{2\pi i\omega_{\hat{\mathbb{F}}}}, k \cdot e^{2\pi i\rho_{\hat{\mathbb{F}}}}\right)$.

Definition 2.4 (Yaqoob et al. 2019)

A CIFG defined on a universal set X is a pair $\mathbb{G} = (\mathcal{P}, \mathcal{Q})$ where $\mathcal{P}$ is an CIFS on vertex set V and $\mathcal{Q}$ is a CIFS on edge set $E \subseteq V \times V$ such that

$$\begin{cases} r_{\mathcal{Q}}(u,v) \cdot e^{i\omega_{\mathcal{Q}}(u,v)} \leq \min\{r_{\mathcal{P}}(u), r_{\mathcal{P}}(v)\} \cdot e^{i\min\{\omega_{\mathcal{P}}(u), \omega_{\mathcal{P}}(v)\}} \\ r_{\mathcal{Q}}(u,v) \cdot e^{i\rho_{\mathcal{Q}}(u,v)} \leq \max\{k_{\mathcal{P}}(u), k_{\mathcal{P}}(v)\} \cdot e^{i\min\{\rho_{\mathcal{P}}(u), \rho_{\mathcal{P}}(v)\}} \end{cases}$$

where $r_{\mathcal{P}}(u), k_{\mathcal{P}}(u) \in [0,1]$, $0 \leq r_{\mathcal{P}}(u) + k_{\mathcal{P}}(u) \leq 1$; $\omega_{\mathcal{P}}(u), \rho_{\mathcal{P}}(u) \in [0, 2\pi]$, such that $0 \leq \omega_{\mathcal{P}}(u) + \rho_{\mathcal{P}}(u) \leq 2\pi$, for all $u, v \in V$.

Definition 2.5 (Akram and Naz 2019)

A CPyFG defined on a universal set X is a pair $\mathbb{G} = (\mathcal{P}, \mathcal{Q})$ where $\mathcal{P}$ is an CPyFS on vertex set V and $\mathcal{Q}$ is a CPyFS on edge set $E \subseteq V \times V$ such that

$$\begin{cases} r_{\mathcal{Q}}(u,v) \cdot e^{i\omega_{\mathcal{Q}}(u,v)} \leq \min\{r_{\mathcal{P}}(u), r_{\mathcal{P}}(v)\} \cdot e^{i\min\{\omega_{\mathcal{P}}(u), \omega_{\mathcal{P}}(v)\}} \\ r_{\mathcal{Q}}(u,v) \cdot e^{i\rho_{\mathcal{Q}}(u,v)} \leq \max\{k_{\mathcal{P}}(u), k_{\mathcal{P}}(v)\} \cdot e^{i\min\{\rho_{\mathcal{P}}(u), \rho_{\mathcal{P}}(v)\}} \end{cases}$$

Where $r_{\mathcal{P}}(u), k_{\mathcal{P}}(u) \in [0,1]$, $0 \leq r_{\mathcal{P}}^2(u) + k_{\mathcal{P}}^2(v) \leq 1$; $\omega_{\mathcal{P}}(u), \rho_{\mathcal{P}}(u) \in [0, 2\pi]$, such that $0 \leq \omega_{\mathcal{P}}^2(u) + \rho_{\mathcal{P}}^2(u) \leq 2\pi$, for all $u, v \in V$.

5.3 Proposed CCFG

In this section, the innovative concept CCFG are introduced and some related properties are investigated.

Definition 3.1 A CFFG defined on a universal set X is a pair $\mathbb{G} = (\mathcal{P}, \mathcal{Q})$ where $\mathcal{P}$ is an CFFS on vertex set V and $\mathcal{Q}$ is a CFFS on edge set $E \subseteq V \times V$ such that

$$\begin{cases} r_{\mathcal{Q}}(u,v) \cdot e^{i\omega_{\mathcal{Q}}(u,v)} \leq \min\{r_{\mathcal{P}}(u), r_{\mathcal{P}}(v)\} \cdot e^{i\min\{\omega_{\mathcal{P}}(u), \omega_{\mathcal{P}}(v)\}} \\ r_{\mathcal{Q}}(u,v) \cdot e^{i\rho_{\mathcal{Q}}(u,v)} \leq \max\{k_{\mathcal{P}}(u), k_{\mathcal{P}}(v)\} \cdot e^{i\min\{\rho_{\mathcal{P}}(u), \rho_{\mathcal{P}}(v)\}} \end{cases}$$

where $r_{\mathcal{P}}(u), k_{\mathcal{P}}(u) \in [0,1]$, $0 \leq r_{\mathcal{P}}^3(u) + k_{\mathcal{P}}^3(v) \leq 1$; $\omega_{\mathcal{P}}(u), \rho_{\mathcal{P}}(u) \in [0, 2\pi]$, such that $0 \leq \omega_{\mathcal{P}}^3(u) + \rho_{\mathcal{P}}^3(u) \leq 2\pi$, for all $u, v \in V$.

Definition 3.2 Let $\mathbb{G} = (\mathcal{P}, \mathcal{Q})$ be CFFG, where $\mathcal{P} = \left\{u, r_{\mathcal{P}}(u) \cdot e^{i\omega_{\mathcal{P}}(u)} | u \in V\right\}$ is the vertex set of $\mathbb{G}$ and $\mathcal{Q} = \left\{(u,v), r_{\mathcal{P}}(u,v) \cdot e^{i\omega_{\mathcal{P}}(u,v)} | (u,v) \in E\right\}$ be the edge set of $\mathbb{G}$, then the order of a CFFG is defined by

$$O(\mathbb{G}) = \left(\sum_{u \in V} r_{\mathcal{P}}(u) \cdot e^{\sum_{u \in V} \omega_{\mathcal{P}}(u)}, \sum_{u \in V} k_{\mathcal{P}}(u) \cdot e^{\sum_{u \in V} \rho_{\mathcal{P}}(u)} \right)$$

And the degree of vertex $u \in V$ is defined by

$$D(u) = \left(\sum_{(u,v) \in E} r_{\mathcal{Q}}(u,v) \cdot e^{\sum_{(u,v) \in E} \omega_{\mathcal{Q}}(u,v)}, \sum_{(u,v) \in E} k_{\mathcal{Q}}(u) \cdot e^{\sum_{(u,v) \in V} \rho_{\mathcal{Q}}(u,v)} \right)$$

Definition 3.3 The Cartesian product $\mathbb{G}_1 \times \mathbb{G}_2$ of two CFFG is defined as a pair $\mathbb{G}_1 \times \mathbb{G}_2 = (\mathcal{P}_1 \times \mathcal{P}_2, \mathcal{Q}_1 \times \mathcal{Q}_2)$, such that

1. $r_{\mathcal{P}_1 \times \mathcal{P}_2}(u, v) \cdot e^{i\omega_{\mathcal{P}_1 \times \mathcal{P}_2}(u,v)} = \min\{r_{\mathcal{P}_1}(u), r_{\mathcal{P}_2}(v)\} \cdot e^{i \min\{\omega_{\mathcal{P}_1}(u), \omega_{\mathcal{P}_2}(v)\}}$
$k_{\mathcal{P}_1 \times \mathcal{P}_2}(u, v) \cdot e^{i\rho_{\mathcal{P}_1 \times \mathcal{P}_2}(u,v)} = \max\{k_{\mathcal{P}_1}(u), k_{\mathcal{P}_2}(v)\} \cdot e^{i \max\{\rho_{\mathcal{P}_1}(u), \rho_{\mathcal{P}_2}(v)\}}$ for all $u, v \in \mathrm{V}$.

2. $r_{\mathcal{Q}_1 \times \mathcal{Q}_2}\langle(u, u_2), (u, v_2)\rangle \cdot e^{i\omega_{\mathcal{Q}_1 \times \mathcal{Q}_2}\langle(u,u_2),(u,v_2)\rangle} = \min\{r_{\mathcal{P}_1}(u), r_{\mathcal{Q}_2}(u_2, v_2)\} \cdot e^{i \min\{\omega_{\mathcal{P}_1}(u), \omega_{\mathcal{Q}_2}(u_2,v_2)\}}$

$$k_{\mathcal{Q}_1 \times \mathcal{Q}_2}\langle(u, u_2), (u, v_2)\rangle \cdot e^{i\rho_{\mathcal{Q}_1 \times \mathcal{Q}_2}\langle(u,u_2),(u,v_2)\rangle}$$
$$= \max\{k_{\mathcal{P}_1}(u), k_{\mathcal{Q}_2}(u_2, v_2)\} \cdot e^{imax\{\rho_{\mathcal{P}_1}(u), \rho_{\mathcal{Q}_2}(u_2,v_2)\}}$$
for all $u \in \mathrm{V}_1$, $(u_2, v_2) \in \mathrm{E}_2$.

3. $r_{\mathcal{Q}_1 \times \mathcal{Q}_2}\langle(u_1, w), (v_1, w)\rangle \cdot e^{i\omega_{\mathcal{Q}_1 \times \mathcal{Q}_2}\langle(u_1,w),(v_1,w)\rangle} = \min\{r_{\mathcal{Q}_1}(u_{1,}, v_1), r_{\mathcal{P}_2}(w)\} \cdot e^{i \min\{\omega_{\mathcal{Q}_1}(u_{1,},v_1), \omega_{\mathcal{P}_2}(w)\}}$

$$k_{\mathcal{Q}_1 \times \mathcal{Q}_2}\langle(u_1, w), (v_1, w)\rangle \cdot e^{i\rho_{\mathcal{Q}_1 \times \mathcal{Q}_2}\langle(u_1,w),(v_1,w)\rangle}$$
$$= \max\{k_{\mathcal{Q}_1}(u_{1,}, v_1), k_{\mathcal{P}_2}(w)\} \cdot e^{i \max\{\omega_{\mathcal{Q}_1}(u_{1,},v_1), \omega_{\mathcal{P}_2}(w)\}}$$
for all $w \in \mathrm{V}_2$, $(u_1, v_1) \in \mathrm{E}_1$.

5.3.1 Basic Operations of CCFG

In this section, some basic operations of CCFG are introduced and some related properties are investigated.

Definition 3.4 The complement of a CFFG $\mathbb{G} = (\mathcal{P}, \mathrm{Q})$ on an underlying graph $G = (V, E)$ is a CFFG $\overline{\mathbb{G}} = (\overline{\mathcal{P}}, \overline{\mathrm{Q}})$ defined by

1. $\overline{r_{\mathcal{P}}(u) \cdot e^{i\omega_{\mathcal{P}}(u)}} = r_{\mathcal{P}}(u) \cdot e^{i\omega_{\mathcal{P}}(u)}$

and

$$\overline{k_{\mathcal{P}}(u) \cdot e^{i\omega_{\mathcal{P}}(u)}} = r_{\mathcal{P}}(u) \cdot e^{i\rho_{\mathcal{P}}(u)}$$

.

2. $$\overline{r_{\mathcal{Q}}(u, v) \cdot e^{i\omega_{\mathcal{Q}}(u,v)}} = \begin{cases} \min\{r_{\mathcal{P}}(u), r_{\mathcal{P}}(v)\} \cdot e^{i \min\{\omega_{\mathcal{P}}(u), \omega_{\mathcal{P}}(v)\}} & \text{if } r_{\mathcal{Q}}(u, v) \cdot e^{i\omega_{\mathcal{Q}}(u,v)} = 0 \\ \min\{r_{\mathcal{P}}(u), r_{\mathcal{P}}(v)\} \cdot e^{i\min\{\omega_{\mathcal{P}}(u), \omega_{\mathcal{P}}(v)\}} - r_{\mathcal{Q}}(u, v) \cdot e^{i\omega_{\mathcal{Q}}(u,v)} & if\, 0 \le r_{\mathcal{Q}}(u, v) \cdot e^{i\omega_{\mathcal{Q}}(u,v)} \le 1 \end{cases}$$

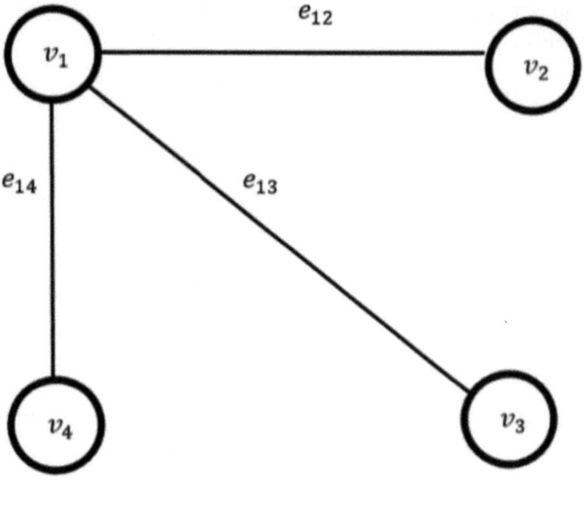

Fig. 5.1 CFFG G

3. $\overline{k_{\mathcal{Q}}(u,v)\cdot e^{i\rho_{\mathcal{Q}}(u,v)}}$

$$= \begin{cases} \max\{k_{\mathcal{P}}(u), k_{\mathcal{P}}(v)\}\cdot e^{i\max\{\rho_{\mathcal{P}}(u),\rho_{\mathcal{P}}(v)\}} & \text{if } k_{\mathcal{Q}}(u,v)\cdot e^{i\rho_{\mathcal{Q}}(u,v)} = 0 \\ \max\{k_{\mathcal{P}}(u), k_{\mathcal{P}}(v)\}\cdot e^{i\max\{\rho_{\mathcal{P}}(u),\rho_{\mathcal{P}}(v)\}} - k_{\mathcal{Q}}(u,v)\cdot e^{i\rho_{\mathcal{Q}}(u,v)} & \text{if } 0 \le k_{\mathcal{Q}}(u,v)\cdot e^{i\rho_{\mathcal{Q}}(u,v)} \le 1 \end{cases}$$

Example 1 Consider a graph $G = (V, E)$ where $V =$ is a vertex set and $E = \{e_1e_2, e_1e_3, e_1e_4\}$ is the edge set of G. Let $\mathbb{G} = (\mathcal{P}, \mathcal{Q})$ be a CFFG on G as given in Fig. 5.1. Table 5.1 provides vertex set $\mathcal{P}$ of CFFG, Table 5.2 provides edge set $\mathcal{Q}$ of CFFG, and Table 5.3 provides edge set of complement CFFG. Figure 5.2, Tables 5.4, 5.5, 5.6 and 5.7.

Theorem 1 The complement of a complement of CFFG is CFFG itself, that is, $\overline{\overline{\mathbb{G}}} = \mathbb{G}$.

Definition 3.5 The union of two CFFGs $\mathbb{G}_1 = (\mathcal{P}_1, \mathcal{Q}_1)$ and $\mathbb{G}_2 = (\mathcal{P}_2, \mathcal{Q}_2)$ respectively is defined as follows:

Table 5.3 SPP under various fuzzy environment

References	Year	Significance influences
Gani and Jabarulla (2010)	2010	SPP with intuitionistic fuzzy variables
Baba (2013)	2013	SPP with intuitionistic fuzzy variables
Khorsandi et al. (2016)	2016	SPP with fuzzy variables
Jan et al. (2019)	2019	SPP with interval-valued Pythagorean fuzzy variables
Proposed		SPP with complex Fermatean fuzzy variables

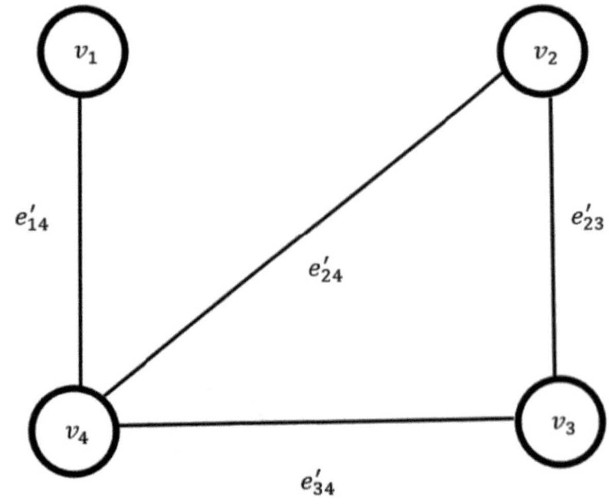

Fig. 5.2 Complement of CFFG $\overline{\mathbb{G}}$

Table 5.4 Vertex set $\mathcal{P}$ of CFFG

Vertex	CFFS Vertex value
v_1	$\langle 0.6e^{i2\pi(0.7)}, 0.5e^{i2\pi(0.6)} \rangle$
v_2	$\langle 0.5e^{i2\pi(0.7)}, 0.3e^{i2\pi(0.8)} \rangle$
v_3	$\langle 0.3e^{i2\pi(0.8)}, 0.2e^{i2\pi(0.5)} \rangle$
v_4	$\langle 0.8e^{i2\pi(0.6)}, 0.6e^{i2\pi(0.4)} \rangle$

Table 5.5 Edge set $\mathcal{Q}$ of CFFG

Edge	CFFS edge value
e_{12}	$\langle 0.5e^{i2\pi(0.7)}, 0.3e^{i2\pi(0.6)} \rangle$
e_{13}	$\langle 0.3e^{i2\pi(0.7)}, 0.2e^{i2\pi(0.5)} \rangle$
e_{14}	$\langle 0.4e^{i2\pi(0.8)}, 0.4e^{i2\pi(0.5)} \rangle$

Table 5.6 Vertex set $\overline{\mathcal{P}}$, of CFFG

Vertex	CFFS vertex value
v_1	$\langle 0.6e^{i2\pi(0.7)}, 0.5e^{i2\pi(0.6)} \rangle$
v_2	$\langle 0.5e^{i2\pi(0.7)}, 0.3e^{i2\pi(0.8)} \rangle$
v_3	$\langle 0.3e^{i2\pi(0.8)}, 0.2e^{i2\pi(0.5)} \rangle$
v_4	$\langle 0.8e^{i2\pi(0.6)}, 0.6e^{i2\pi(0.4)} \rangle$

Table 5.7 Edge set $\overline{\mathcal{Q}}$ of complement CFFG

Vertex	CFFS Vertex value
$e_{14'}$	$\langle 0.2e^{i2\pi(0.6)}, 0.1e^{i2\pi(0.4)} \rangle$
$e_{23'}$	$\langle 0.4e^{i2\pi(0.5)}, 0.3e^{i2\pi(0.5)} \rangle$
$e_{24'}$	$\langle 0.5e^{i2\pi(0.6)}, 0.3e^{i2\pi(0.4)} \rangle$
$e_{34'}$	$\langle 0.4e^{i2\pi(0.5)}, 0.7e^{i2\pi(0.5)} \rangle$

i.
$$\left(r_{\mathcal{P}_1} \cup r_{\mathcal{P}_2}\right)(u) \cdot e^{i(\omega_{\mathcal{P}_1} \cup \omega_{\mathcal{P}_2})(u)} = \begin{cases} r_{\mathcal{P}_1}(u) \cdot e^{i(\omega_{\mathcal{P}_1})(u)}, & \text{if } u \in \mathcal{P}_1 - \mathcal{P}_2 \\ r_{\mathcal{P}_2}(u) \cdot e^{i(\omega_{\mathcal{P}_2})(u)}, & \text{if } u \in \mathcal{P}_2 - \mathcal{P}_1 \\ \max\left(r_{\mathcal{P}_1}(u), r_{\mathcal{P}_2}(u)\right) \cdot e^{imax\left(\omega_{\mathcal{P}_1}(u), \omega_{\mathcal{P}_2}(u)\right)}, & \text{if } u \in \mathcal{P}_1 \cap \mathcal{P}_2 \end{cases}$$

ii.
$$\left(k_{\mathcal{P}_1} \cup k_{\mathcal{P}_2}\right)(u) \cdot e^{i(\rho_{\mathcal{P}_1} \cup \rho_{\mathcal{P}_2})(u)} = \begin{cases} k_{\mathcal{P}_1}(u) \cdot e^{i(\rho_{\mathcal{P}_1})(u)}, \, ifu \in \mathcal{P}_1 - \mathcal{P}_2 \\ k_{\mathcal{P}_2}(u) \cdot e^{i(\rho_{\mathcal{P}_2})(u)}, \, ifu \in \mathcal{P}_2 - \mathcal{P}_1 \\ \min\left(k_{\mathcal{P}_1}(u), k_{\mathcal{P}_2}(u)\right) \cdot e^{imin\left(\rho_{\mathcal{P}_1}(u), \rho_{\mathcal{P}_2}(u)\right)}, \, ifu \in \mathcal{P}_1 \cap \mathcal{P}_2 \end{cases}$$

iii.
$$\left(r_{Q_1} \cup r_{Q_2}\right)(u, v) \cdot e^{i(\omega_{Q_1} \cup \omega_{Q_2})(u,v)} = \begin{cases} r_{Q_1}(u, v) \cdot e^{i(\omega_{Q_1})(u,v)}, & \text{if } (u, v) \in Q_1 - Q_2 \\ r_{Q_2}(u, v) \cdot e^{i(\omega_2)(u,v)}, & \text{if } (u, v) \in Q_2 - Q_1 \\ \max\left(r_{Q_1}(u, v), r_{Q_2}(u, v)\right) \cdot e^{imax\left(\omega_{Q_1}(u,v), \omega_{Q_2}(u,v)\right)}, & \text{if } (u, v) \in Q_1 \cap Q_2 \end{cases}$$

iv.
$$\left(k_{Q_1} \cup k_{Q_2}\right)(u, v) \cdot e^{i(\rho_{Q_1} \cup \rho_{Q_2})(u,v)} = \begin{cases} k_{Q_1}(u, v) \cdot e^{i(\rho_{Q_1})(u,v)}, & \text{if } (u, v) \in Q_1 - Q_2 \\ k_{Q_2}(u, v) \cdot e^{i(\rho_{Q_2})(u,v)}, & \text{if } (u, v) \in Q_2 - Q_1 \\ \min\left(k_{Q_1}(u, v), k_{Q_2}(u, v)\right) \cdot e^{i\min\left(\rho_{Q_1}(u,v), \rho_{Q_2}(u,v)\right)}, & \text{if } (u, v) \in Q_1 \cap Q_2 \end{cases}$$

Theorem 2 The join $\mathbb{G}_1 + \mathbb{G}_2$ of $\mathbb{G}_1$ and $\mathbb{G}_2$ is a CFFG of $G_1 + G_2$ if and only if $\mathbb{G}_1$ and $\mathbb{G}_2$ are CFFGs of G_1 and G_2, respectively, where $\mathcal{P}_1 \cap \mathcal{P}_2 = \emptyset$.

Definition 3.6 The ring-sum $\mathbb{G}_1 \oplus \mathbb{G}_2 = (\mathcal{P}_1 \oplus \mathcal{P}_2, \mathrm{Q}_1 \oplus \mathrm{Q}_2)$ of two CFFSs $\mathbb{G}_1 = (\mathcal{P}_1, Q_1)$ and $\mathbb{G}_2 = (\mathcal{P}_2, Q_2)$ of the graph $\mathbb{G}_1$ and $\mathbb{G}_2$ respectively is defined as follows:

i.
$$\begin{cases} \left(r_{\mathcal{P}_1} \oplus r_{\mathcal{P}_2}\right)(u) \cdot e^{i(\omega_{\mathcal{P}_1} \oplus \omega_{\mathcal{P}_2})(u)} = \left(r_{\mathcal{P}_1} \cup r_{\mathcal{P}_2}\right)(u) \cdot e^{i(\omega_{\mathcal{P}_1} \cup \omega_{\mathcal{P}_2})(u)} \\ \left(k_{\mathcal{P}_1} \oplus k_{\mathcal{P}_2}\right)(u) \cdot e^{i(\rho_{\mathcal{P}_1} \oplus \rho_{\mathcal{P}_2})(u)} = \left(k_{\mathcal{P}_1} \cup k_{\mathcal{P}_2}\right)(u) \cdot e^{i(\rho_{\mathcal{P}_1} \cup \rho_{\mathcal{P}_2})(u)} \quad \text{if } u \in \mathcal{P}_1 \cup \mathcal{P}_2 \end{cases}$$

ii.
$$\left(r_{Q_1} \oplus r_{Q_2}\right)(u, v) \cdot e^{i(\omega_{Q_1} \oplus \omega_{Q_2})(u,v)} = \begin{cases} r_{Q_1}(u, v) \cdot e^{i(\omega_{Q_1})(u,v)}, & \text{if } (u, v) \in Q_1 - Q_2 \\ r_{Q_2}(u, v) \cdot e^{i(\omega_2)(u,v)}, & \text{if } (u, v) \in Q_2 - Q_1 \\ 0, & \text{if } (u, v) \in Q_1 \cap Q_2 \end{cases}$$

iii.
$$\left(k_{Q_1} \oplus k_{Q_2}\right)(u, v) \cdot e^{i(\rho_{Q_1} \oplus \rho_{Q_2})(u,v)} = \begin{cases} k_{Q_1}(u, v) \cdot e^{i(\rho_{Q_1})(u,v)}, & \text{if } (u, v) \in Q_1 - Q_2 \\ k_{Q_2}(u, v) \cdot e^{i(\rho_{Q_2})(u,v)}, & \text{if } (u, v) \in Q_2 - Q_1 \\ 0, & \text{if } (u, v) \in Q_1 \cap Q_2 \end{cases}$$

Remark If $\mathbb{G}_1 = (\mathcal{P}_1, Q_1)$ and $\mathbb{G}_2 = (\mathcal{P}_2, Q_2)$ are the CFFGs, then $\mathbb{G}_1 \oplus \mathbb{G}_2$ is the CFFG,

Definition 3.7 Let $\mathbb{G}_1$ and $\mathbb{G}_2$ are CFFGs of G_1 and G_2, respectively. The join $\mathbb{G}_1 + \mathbb{G}_2 = (\mathcal{P}_1 + \mathcal{P}_2, \mathcal{Q}_1 + \mathcal{Q}_2)$ of $\mathbb{G}_1$ and $\mathbb{G}_2$ is defined as:

i. $$\begin{cases} (r_{\mathcal{P}_1} + r_{\mathcal{P}_2})(u) \cdot e^{i(\omega_{\mathcal{P}_1} + \omega_{\mathcal{P}_2})(u)} = (r_{\mathcal{P}_1} \cup r_{\mathcal{P}_2})(u) \cdot e^{i(\omega_{\mathcal{P}_1} \cup \omega_{\mathcal{P}_2})(u)} \\ (k_{\mathcal{P}_1} + k_{\mathcal{P}_2})(u) \cdot e^{i(\rho_{\mathcal{P}_1} + \rho_{\mathcal{P}_2})(u)} = (k_{\mathcal{P}_1} \cup k_{\mathcal{P}_2})(u) \cdot e^{i(\rho_{\mathcal{P}_1} \cup \rho_{\mathcal{P}_2})(u)}, \quad \text{if } u \in \mathcal{P}_1 \cup \mathcal{P}_2 \end{cases}$$

ii. $$\begin{cases} (r_{\mathcal{Q}_1} \oplus r_{\mathcal{Q}_2})(u, v) \cdot e^{i(\omega_{\mathcal{Q}_1} \oplus \omega_{\mathcal{Q}_2})(u,v)} = (r_{\mathcal{Q}_1} \cup r_{\mathcal{Q}_2})(u, v) \cdot e^{i(\omega_{\mathcal{Q}_1} \cup \omega_{\mathcal{Q}_2})(u,v)} \\ (k_{\mathcal{Q}_1} \oplus k_{\mathcal{Q}_2})(u, v) \cdot e^{i(\rho_{\mathcal{Q}_1} \oplus \rho_{\mathcal{Q}_2})(u,v)} = (k_{\mathcal{Q}_1} \cup k_{\mathcal{Q}_2})(u, v) \cdot e^{i(\rho_{\mathcal{Q}_1} \cup \rho_{\mathcal{Q}_2})(u,v)}, \quad \text{if } (u, v) \in \mathcal{Q}_1 \cup \mathcal{Q}_2 \end{cases}$$

iii. $$\begin{cases} (r_{\mathcal{Q}_1} \oplus r_{\mathcal{Q}_2})(u, v) \cdot e^{i(\omega_{\mathcal{Q}_1} \oplus \omega_{\mathcal{Q}_2})(u,v)} = \min(r_{\mathcal{P}_1}(u), r_{\mathcal{P}_2}(v)) \cdot e^{i \min(\omega_{\mathcal{P}_1}(u), \omega_{\mathcal{P}_2}(v))} \\ (k_{\mathcal{Q}_1} \oplus k_{\mathcal{Q}_2})(u, v) \cdot e^{i(\rho_{\mathcal{Q}_1} \oplus \rho_{\mathcal{Q}_2})(u,v)} = \max(k_{\mathcal{P}_1}(u), k_{\mathcal{P}_2}(v)) \cdot e^{i\min(\rho_{\mathcal{P}_1}(u), \rho_{\mathcal{P}_2}(v))}, \quad \text{if } (u, v) \in \mathcal{Q}\prime \end{cases}$$

where $\mathcal{Q}\prime$ is the set of all edges joining the vertices of $\mathcal{P}_1$ and $\mathcal{P}_2$, $\mathcal{P}_1 \cap \mathcal{P}_2 = \emptyset$.

Theorem 3 The join $\mathbb{G}_1 + \mathbb{G}_2$ of $\mathbb{G}_1$ and $\mathbb{G}_2$ is a CFFG of $G_1 + G_2$ iff $\mathbb{G}_1$ and $\mathbb{G}_2$ are CFFGs of G_1 and G_2, where $\mathcal{P}_1 \cap \mathcal{P}_2 = \emptyset$.

Definition 3.8 The total degree of a vertex $v \in V$ in a CFFG $\mathbb{G}$ is denoted by $\text{TD}(u)$ and is defined by

$$\text{TD}(u) = \left(\text{TD}_r e^{i\omega(u)}, \text{TD}_k e^{i\rho(u)}\right), \text{ where}$$

$$\text{TD}_r e^{i\omega(u)} = \sum_{(u,v) \neq u \in V} r_{\mathcal{Q}}(u, v) \cdot e^{\sum_{(u,v) \neq u \in V} \omega_{\mathcal{Q}}(u,v)} + r_{\mathcal{P}}(u) \cdot e^{i\omega_{\mathcal{P}}(u)}$$

$$\text{TD}_k e^{i\rho(u)} = \sum_{(u,v) \neq u \in V} k_{\mathcal{Q}}(u, v) \cdot e^{\sum_{(u,v) \neq u \in V} \rho_{\mathcal{Q}}(u,v)} + k_{\mathcal{P}}(u) \cdot e^{i\rho_{\mathcal{P}}(u)}$$

Definition 3.9 Let $\mathbb{G}_1$ and $\mathbb{G}_2$ be two CFFGs. For any vertex $u \in \mathcal{P}_1 \cup \mathcal{P}_2$, there are three cases to consider.

Case 1: Either $u \in \mathcal{P}_1 - \mathcal{P}_2$ or $u \in \mathcal{P}_2 - \mathcal{P}_1$. Then, no edge incident at u lies in $\mathcal{Q}_1 \cup \mathcal{Q}_2$. Thus, for $u \in \mathcal{P}_1 - \mathcal{P}_2$

$$(D_{re^{i\omega}})_{\mathbb{G}_1 \cup \mathbb{G}_2}(u) = \sum_{(u,v) \in \mathrm{Q}_1} r_{\mathrm{Q}_1}(u, v) \cdot e^{\sum_{(u,v) \in \mathrm{Q}_1} \omega_{\mathrm{Q}_1}(u,v)} = (D_{re^{i\omega}})_{\mathbb{G}_1}(u)$$

$$(D_{ke^{i\rho}})_{\mathbb{G}_1 \cup \mathbb{G}_2}(u) = \sum_{(u,v) \in \mathrm{Q}_1} k_{\mathcal{Q}_1}(u, v) \cdot e^{\sum_{(u,v) \in \mathrm{Q}_1} \rho_{\mathrm{Q}_1}(u,v)} = (D_{ke^{i\rho}})_{\mathbb{G}_1}(u)$$

$$(TD_{re^{i\omega}})_{\mathbb{G}_1 \cup \mathbb{G}_2}(u) = (TD_{re^{i\omega}})_{\mathbb{G}_1}(u); \ (TD_{ke^{i\rho}})_{\mathbb{G}_1 \cup \mathbb{G}_2}(u) = (TD_{ke^{i\rho}})_{\mathbb{G}_1}(u)$$

Case 2: Either $u \in \mathcal{P}_1 \cap \mathcal{P}_2$ but no edge incident at $u \in \mathcal{Q}_1 \cap \mathcal{Q}_2$. Then, any edge incident at u is either in $\mathcal{Q}_1 - \mathcal{Q}_2$ or $\mathcal{Q}_2 - \mathcal{Q}_1$.

$$(D_{re^{i\omega}})_{\mathbb{G}_1 \cup \mathbb{G}_2}(u) = (D_{re^{i\omega}})_{\mathbb{G}_1}(u) + (D_{r\omega})_{\mathbb{G}_2}(u)$$

$$(D_{ke^{i\omega}})_{\mathbb{G}_1 \cup \mathbb{G}_2}(u) = (D_{ke^{i\rho}})_{\mathbb{G}_1}(u) + (D_{ke^{i\rho}})_{\mathbb{G}_2}(u)$$

$$(\mathrm{TD}_{re^{i\omega}})_{\mathbb{G}_1 \cup \mathbb{G}_2}(u) = (\mathrm{TD}_{re^{i\omega}})_{\mathbb{G}_1}(u) + (\mathrm{TD}_{re^{i\omega}})_{\mathbb{G}_2}(u) - \min\big(r_{\mathcal{P}_1}(u), r_{\mathcal{P}_2}(u)\big) \cdot e^{i\min(\omega_{\mathcal{P}_1}(u), \omega_{\mathcal{P}_2}(u))}$$

$$(\mathrm{TD}_{ke^{i\rho}})_{\mathbb{G}_1 \cup \mathbb{G}_2}(u) = (\mathrm{TD}_{ke^{i\rho}})_{\mathbb{G}_1}(u) + (\mathrm{TD}_{ke^{i\rho}})_{\mathbb{G}_2}(u) - \max\big(r_{\mathcal{P}_1}(u), r_{\mathcal{P}_2}(u)\big) \cdot e^{i\max(\omega_{\mathcal{P}_1}(u), \omega_{\mathcal{P}_2}(u))}$$

Case 3: $u \in \mathcal{P}_1 \cap \mathcal{P}_2$ and some edges incident at u are in $\mathcal{Q}_1 \cap \mathcal{Q}_2$.

$$(D_{re^{i\omega}})_{\mathbb{G}_1 \cup \mathbb{G}_2}(u) = (D_{re^{i\omega}})_{\mathbb{G}_1}(u) + (D_{re^{i\omega}})_{\mathbb{G}_2}(u) - \sum_{(u,v)\in \mathcal{Q}_1 \cap \mathcal{Q}_2} \min\big(r_{\mathcal{Q}_1}(u,v), r_{\mathcal{Q}_2}(u,v)\big) \cdot e^{i\sum_{(u,v)\in \mathcal{Q}_1 \cap \mathcal{Q}_2} \min(\omega_{\mathcal{Q}_1}(u,v), \omega_{\mathcal{Q}_2}(u,v))}$$

$$(D_{ke^{i\omega}})_{\mathbb{G}_1 \cup \mathbb{G}_2}(u) = (D_{ke^{i\rho}})_{\mathbb{G}_1}(u) + (D_{ke^{i\rho}})_{\mathbb{G}_2}(u) - \sum_{(u,v)\in \mathcal{Q}_1 \cap \mathcal{Q}_2} \max\big(k_{\mathcal{Q}_1}(u,v), k_{\mathcal{Q}_2}(u,v)\big) \cdot e^{i\sum_{(u,v)\in \mathcal{Q}_1 \cap \mathcal{Q}_2} \max(\rho_{\mathcal{Q}_1}(u,v), \rho_{\mathcal{Q}_2}(u,v))}$$

$$(\mathrm{TD}_{re^{i\omega}})_{\mathbb{G}_1 \cup \mathbb{G}_2}(u) = (\mathrm{TD}_{re^{i\omega}})_{\mathbb{G}_1}(u) + (\mathrm{TD}_{re^{i\omega}})_{\mathbb{G}_2}(u) - \sum_{(u,v)\in \mathcal{Q}_1 \cap \mathcal{Q}_2} \min\big(r_{\mathcal{Q}_1}(u,v), r_{\mathcal{Q}_2}(u,v)\big) \cdot e^{i\sum_{(u,v)\in \mathcal{Q}_1 \cap \mathcal{Q}_2} \min(\omega_{\mathcal{Q}_1}(u,v), \omega_{\mathcal{Q}_2}(u,v))}$$

$$(\mathrm{TD}_{ke^{i\rho}})_{\mathbb{G}_1 \cup \mathbb{G}_2}(u) = (\mathrm{TD}_{ke^{i\rho}})_{\mathbb{G}_1}(u) + (\mathrm{TD}_{ke^{i\rho}})_{\mathbb{G}_2}(u) - \sum_{(u,v)\in \mathcal{Q}_1 \cap \mathcal{Q}_2} \max\big(k_{\mathcal{Q}_1}(u,v), k_{\mathcal{Q}_2}(u,v)\big) \cdot e^{i\sum_{(u,v)\in \mathcal{Q}_1 \cap \mathcal{Q}_2} \max(\rho_{\mathcal{Q}_1}(u,v), \rho_{\mathcal{Q}_2}(u,v))}$$

Example 2 Consider two CFFGs $\mathbb{G}_1 = (\mathcal{P}_1, \mathcal{Q}_1)$ and $\mathbb{G}_2 = (\mathcal{P}_2, \mathcal{Q}_2)$ on $\mathcal{P}_1 = \{v_1, v_2, v_3\}$ and $\mathcal{P}_2 = \{v_1, v_2, v_3, v_4\}$ respectively as shown in Figs. 5.3 and 5.4. Tables 5.8, 5.9, 5.10, 5.11, 5.12, and 5.13.

Using Definition 3.2, clearly $v_4 \in \mathcal{P}_2 - \mathcal{P}_1$ and $v_1, v_2, v_3 \in \mathcal{P}_1 \cap \mathcal{P}_2$. Figure 5.5 shows CFFG of $\mathbb{G}_1 \cup \mathbb{G}_2$.

Definition 3.10 Thirunavukarasu et al. 2016)

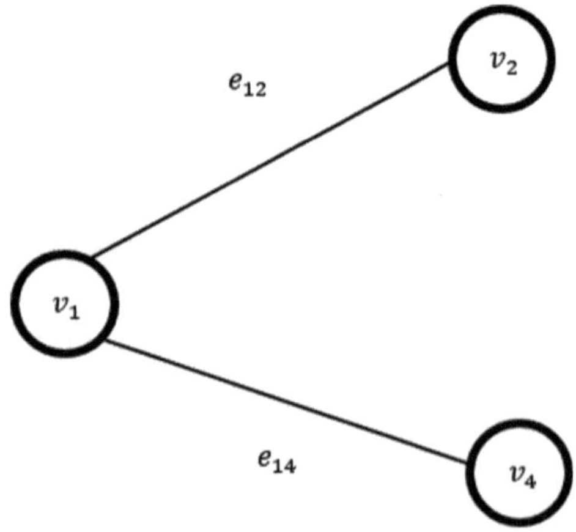

Fig. 5.3 CFFG $\mathbb{G}_1$

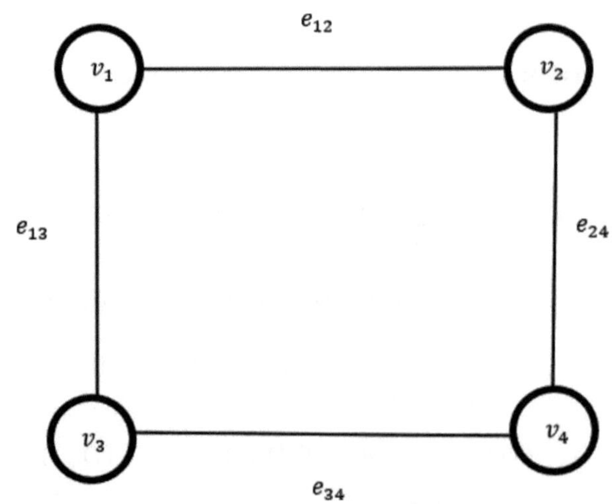

Fig. 5.4 CFFG $\mathbb{G}_1$

Table 5.8 Vertex set $\mathcal{P}_1$ of CFFG $\mathbb{G}_1$

Vertex	CFFS vertex value
v_1	$\langle 0.5e^{i2\pi(0.7)}, 0.8e^{i2\pi(0.6)} \rangle$
v_2	$\langle 0.8e^{i2\pi(0.3)}, 0.6e^{i2\pi(0.9)} \rangle$
v_3	$\langle 0.3e^{i2\pi(0.8)}, 0.2e^{i2\pi(0.5)} \rangle$

Table 5.9 Edge set $\mathcal{Q}_1$ of CFFG $\mathbb{G}_1$

Edge	CFFS edge value
e_{12}	$\langle 0.6e^{i2\pi(0.4)}, 0.6e^{i2\pi(0.7)} \rangle$
e_{14}	$\langle 0.4e^{i2\pi(0.8)}, 0.2e^{i2\pi(0.8)} \rangle$

Table 5.10 Vertex set $\mathcal{P}_2$ of CFFG $\mathbb{G}_2$

Vertex	CFFS vertex value
v_1	$\langle 0.7e^{i2\pi(0.5)}, 0.6e^{i2\pi(0.3)} \rangle$
v_2	$\langle 0.8e^{i2\pi(0.4)}, 0.5e^{i2\pi(0.8)} \rangle$
v_3	$\langle 0.6e^{i2\pi(0.8)}, 0.8e^{i2\pi(0.7)} \rangle$
v_4	$\langle 0.3e^{i2\pi(0.6)}, 0.6e^{i2\pi(0.9)} \rangle$

Table 5.11 Edge set Q_2 of CFFG $\mathbb{G}_2$

Edge	CFFS edge value
e_{12}	$\langle 0.4e^{i2\pi(0.8)}, 0.5e^{i2\pi(0.7)} \rangle$
e_{13}	$\langle 0.6e^{i2\pi(0.7)}, 0.3e^{i2\pi(0.6)} \rangle$
e_{24}	$\langle 0.5e^{i2\pi(0.4)}, 0.8e^{i2\pi(0.7)} \rangle$
e_{34}	$\langle 0.7e^{i2\pi(0.5)}, 0.4e^{i2\pi(0.5)} \rangle$

Table 5.12 Vertex set of CFFG $\mathbb{G}_1 \cup \mathbb{G}_2$

Vertex	CFFS vertex value
v_1	$\langle 0.6e^{i2\pi(0.7)}, 0.5e^{i2\pi(0.6)} \rangle$
v_2	$\langle 0.5e^{i2\pi(0.7)}, 0.3e^{i2\pi(0.8)} \rangle$
v_3	$\langle 0.3e^{i2\pi(0.8)}, 0.2e^{i2\pi(0.5)} \rangle$

Table 5.13 Edge set of CFFG $\mathbb{G}_1 \cup \mathbb{G}_2$

Edge	CFFS edge value
e_{12}	$\langle 0.5e^{i2\pi(0.7)}, 0.3e^{i2\pi(0.6)} \rangle$
e_{13}	$\langle 0.3e^{i2\pi(0.7)}, 0.2e^{i2\pi(0.5)} \rangle$
e_{24}	$\langle 0.5e^{i2\pi(0.7)}, 0.3e^{i2\pi(0.6)} \rangle$
e_{34}	$\langle 0.3e^{i2\pi(0.7)}, 0.2e^{i2\pi(0.5)} \rangle$

Fig. 5.5 CFFG $\mathbb{G}_1 \cup \mathbb{G}_2$

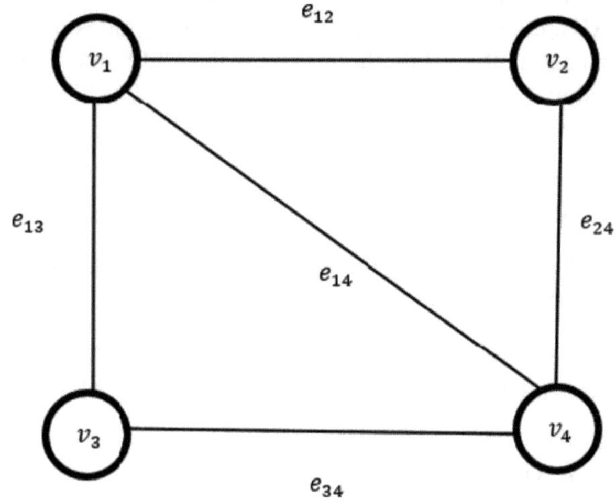

Let $\mathbb{A} = \left(r \cdot e^{i2\pi\omega_r}, k \cdot e^{i2\pi\rho_k}\right)$, $\mathbb{A}_1 = (r_1 \cdot e^{i2\pi\omega_{r_1}}, k_1 \cdot e^{i2\pi\rho_{k_1}})$ and $\mathbb{A}_2 = (r_2 \cdot e^{i2\pi\omega_{r_2}}, k_2 \cdot e^{i2\pi\rho_{k_2}})$ be three CFF and $\lambda > 0$. Then, the operations rules are defined as follows:

$$\mathbb{A}_1 \oplus \mathbb{A}_2 = \left\langle \sqrt[3]{r_1^3 + r_2^3 - r_1^3 r_2^3} \cdot e^{i2\pi \sqrt[3]{\omega_{r_1}^3 + \omega_{r_2}^3 - \omega_{r_1}^3 \omega_{r_2}^3}}, k_1 k_2 \cdot e^{i2\pi(\rho_{k_1}\rho_{k_2})} \right\rangle$$

$$\mathbb{A}_1 \otimes \mathbb{A}_2 = \left\langle r_1 r_2 \cdot e^{i2\pi(\omega_{r_1}\omega_{r_2})}, \sqrt[3]{k_1^3 + k_2^3 - k_1^3 k_2^3} \cdot e^{i2\pi \sqrt[3]{\rho_{k_1}^3 + \rho_{k_2}^3 - \rho_{k_1}^3 \rho_{k_2}^3}} \right\rangle$$

$$\lambda\mathbb{A} == \left\langle \sqrt[3]{1 - \left(1 - r^3\right)^\lambda} \cdot e^{i2\pi \sqrt[3]{1 - (1 - \omega_r^3)^\lambda}}, k^\lambda \cdot e^{i2\pi(\rho_k^\lambda)} \right\rangle$$

$$\mathbb{A}^\lambda == \left\langle r^\lambda \cdot e^{i2\pi(\omega_r^\lambda)}, \sqrt[3]{1 - \left(1 - k^3\right)^\lambda} \cdot e^{i2\pi \sqrt[3]{1 - (1 - \rho_k^3)^\lambda}} \right\rangle$$

Definition 3.11 A CFFN $\mathbb{A} = \left(r \cdot e^{i2\pi\omega_r}, k \cdot e^{i2\pi\rho_k}\right)$ is said to be zero CFFN if and only if

$$r \cdot e^{i2\pi\omega_r} = 0 \cdot e^0 \text{ and } k \cdot e^{i2\pi\rho_k} = 1 \cdot e^{i2\pi}$$

Definition 3.12 A CFFN $\mathbb{A} = \left(r \cdot e^{i2\pi\omega_r}, k \cdot e^{i2\pi\rho_k}\right)$ is said to be full CFFN if and only if

$$r \cdot e^{i2\pi\omega_r} = 1 \cdot e^{i2\pi} \text{and} k \cdot e^{i2\pi\rho_k} = 0 \cdot e^0$$

Definition 3.13 Thirunavukarasu et al. 2016)

In order to make a comparison between two CFFN, Greenfield et al. (2016) proposed the concept of score function (SF) and accuracy function (AF) in 2021. The SF and AF are applied to compare the complex grades of CFFS. This function shows that greater is the value, the greater is the CFFS and by using this concept, paths or arc length can be ranked. Let $\mathbb{A} = \left(r \cdot e^{i2\pi\omega_r}, k \cdot e^{i2\pi\rho_k}\right)$ be a CFFN, then the SF ($\mathbb{A}$) and AF ($\mathbb{A}$) are defined as follows:

- $\mathrm{SF}(\mathbb{A}) = \frac{1}{2}\left((r^3 - k^3) + (\omega_r^3 - \rho_k^3)\right)$
- $\mathrm{AF}(\mathbb{A}) = \frac{1}{2}\left((r^3 + k^3) + (\omega_r^3 + \rho_k^3)\right)$

Depending upon the SF and AF of CFFNs, the ranking technique for any two CFFNs can be described as:

Definition 3.14 Let $\mathbb{A}_1 = (r_1 \cdot e^{i2\pi\omega_{r_1}}, k_1 \cdot e^{i2\pi\rho_{k_1}})$ and $\mathbb{A}_2 = (r_2 \cdot e^{i2\pi\omega_{r_2}}, k_2 \cdot e^{i2\pi\rho_{k_2}})$ be two CFFNs and the accuracy function be AF ($\mathbb{A}_1$), AF ($\mathbb{A}_2$), then

- If AF ($\mathbb{A}_1$) < AF ($\mathbb{A}_2$), then $\mathbb{A}_1 < \mathbb{A}_2$;
- If AF ($\mathbb{A}_1$) > AF ($\mathbb{A}_2$), then $\mathbb{A}_1 > \mathbb{A}_2$;
- If AF ($\mathbb{A}_1$) = AF ($\mathbb{A}_2$), then $\mathbb{A}_1 = \mathbb{A}_2$.

5.4 SPP Under Complex Environment and Proposed Algorithm

In this section, a modified Dijkstra's algorithm to solve CFFSPP is proposed. The SPP is a fundamental problem in network optimization which involves finding a route between two vertices in an undirected, directed or mixed graph such that the sum of the edge weights has to be minimized. In the paper, we focus on directed CFFG. The directed network can be defined by $\mathbb{G} = (\mathcal{P}, \mathcal{Q})$ where $\mathcal{P}$ denotes the set of N nodes and $\mathcal{Q}$ signifies the set of M directed arcs. Each arc (m–n) $\in \mathcal{P}$ has an associated cost (or length) d_{mn}. Usually, one of the two nodes of the wanted path is a specified source node s and the second one is a specified sink node t. In 1956, Dutch computer scientist E. Dijkstra developed the Dijkstra algorithm or labelling algorithm for solving SPP. Since the path lengths in SPP are represented by CFFN, the arithmetic operations and score function of CFFN are used to carry out the traditional Dijkstra's algorithm to find CFSP. For the arc $m - n$, the distance is represented by $d_{mn} = \left(r_{mn} \cdot e^{i2\pi \omega_{r_{mn}}}, k_{mn} \cdot e^{i2\pi \rho_{k_{mn}}}\right)$ where $r_{mn} \cdot e^{i2\pi \omega_{r_{mn}}}$ denotes the membership degree and $k_{mn} \cdot e^{i2\pi \rho_{k_{mn}}}$ denotes the non-membership degree. The FFSPP goal is to achieve the shortest path between ν_1 and ν_{N} in terms of distance, time, or price. Below are the steps of the modified Dijkstra's algorithm in a complex Fermatean environment.

Step 1

Let the initial and end nodes of destination be ν_1 and ν_{N}.

Step 2

Since the distance from initial node v_1 to itself is zero, we assume $d_1 = \langle 0, 1\rangle$ and is labelled as ((0, 1, −).

Step 3

Using the addition arithmetic operation of CFFN, find

$$d_n = \min\{d_m \oplus d_{mn}\} \quad \text{for } j = 2, 3, \ldots N.$$

Step 4

In the case, if the distance value occurs for a unique $m = r$, then n is marked as $[d_n, r]$. Otherwise, it represents more than one CFFN paths from a node. So, to get the shortest distance among several paths, we use the AF to rank CFFN. That is AF $(\mathbb{A}) = \mathrm{AF}(\mathbb{A}) = \frac{1}{2}\left((r^3 + k^3) + (\omega_r^3 + \rho_k^3)\right)$.

Step 5

The shortest distance value of end node is denoted by d_N and is labelled as $[d_N, t]$.

Step 6

To find the CFFN shortest path (CFFNSP) from initial node to end node which is labelled as $[d_N, t]$, check the label of node t and allow to be $[d_N, s]$. Next, we examine node s label and so forth. Continue these iterations to obtain the origin node.

Step 7

Using step 6 the CFFNSP can be obtained.

5.5 Numerical Example

Now, let us illustrate the proposed algorithm to find the SP under complex environment. Consider a network based on CFFG given in Fig. 5.6 where the distance between the vertices is a CFFN and is described in Table 5.14.

Now we compute the SP by using the proposed algorithm as follows:

Represent the initial node by ν_1 and end node by ν_6 as $\mathrm{N} = 6$. Let $\mathrm{d}_1 = \left\langle 0 \cdot e^0, 1 \cdot e^{i2\pi} \right\rangle = \langle 0, 1\rangle$ and denote the source node as 1 and label it as $(\langle 0, 1\rangle, -)$. The distance d_N for $N = 2, 3, 4, 5, 6$ can be computed as follows:

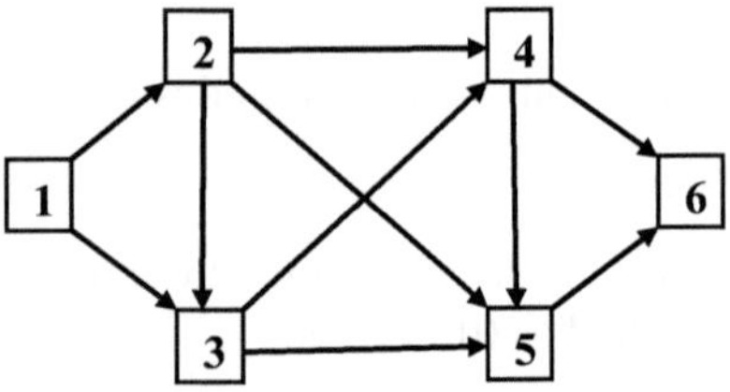

Fig. 5.6 CFF network

Table 5.14 CFF network edges

Edges	Distance
(1, 2)	$\left\langle 0.7 \cdot e^{i2\pi(0.4)}, 0.1 \cdot e^{i2\pi(0.8)} \right\rangle$
(1, 3)	$\left\langle 0.3 \cdot e^{i2\pi(0.6)}, 0.5 \cdot e^{i2\pi(0.7)} \right\rangle$
(2, 3)	$\left\langle 0.1 \cdot e^{i2\pi(0.6)}, 0.8 \cdot e^{i2\pi(0.6)} \right\rangle$
(2, 4)	$\left\langle 0.2 \cdot e^{i2\pi(0.2)}, 0.9 \cdot e^{i2\pi(0.4)} \right\rangle$
(2, 5)	$\left\langle 0.4 \cdot e^{i2\pi(0.8)}, 0.8 \cdot e^{i2\pi(0.2)} \right\rangle$
(3, 4)	$\left\langle 0.3 \cdot e^{i2\pi(0.3)}, 0.8 \cdot e^{i2\pi(0.7)} \right\rangle$
(3, 5)	$\left\langle 0.7 \cdot e^{i2\pi(0.5)}, 0.2 \cdot e^{i2\pi(0.7)} \right\rangle$
(4, 5)	$\left\langle 0.2 \cdot e^{i2\pi(0.4)}, 0.6 \cdot e^{i2\pi(0.9)} \right\rangle$
(4, 6)	$\left\langle 0.5 \cdot e^{i2\pi(0.6)}, 0.4 \cdot e^{i2\pi(0.5)} \right\rangle$
(5,6)	$\left\langle 0.8 \cdot e^{i2\pi(0.6)}, 0.5 \cdot e^{i2\pi(0.3)} \right\rangle$

Iteration 1

As node 1 is the only antecedent of node 2, we calculate d_2 alone. For the path 1–2 ($m = 1$ and $n = 2$), we have

$$\begin{aligned} d_2 &= \min\{d_1 \oplus d_{12}\} = \min(\langle 0 \cdot e^0, 1 \cdot e^{i2\pi} \rangle \oplus \langle 0.7 \cdot e^{i2\pi(0.4)}, 0.1 \cdot e^{i2\pi(0.8)} \rangle \\ &= \langle 0.7 \cdot e^{i2\pi(0.4)}, 0.1 \cdot e^{i2\pi(0.8)} \rangle \end{aligned}$$

Since the minimum value occurs for m = 1 alone, v_2 is labelled as $(\langle 0.7 \cdot e^{i2\pi(0.4)}, 0.1 \cdot e^{i2\pi(0.8)} \rangle, -1)$.

Iteration 2

Now, the predecessor for node 3 is nodes 1 and 2. For the path 1–3, 2–3 ($m = 1, 2$ and $n = 3$), we have

$$\begin{aligned} d_3 &= \min\{d_1 \oplus d_{13}, d_2 \oplus d_{23}\} \\ &= \min\{\langle 0, 1 \rangle \oplus \langle 0.3 \cdot e^{i2\pi(0.6)}, 0.5 \cdot e^{i2\pi(0.7)} \rangle, \langle 0.7 \cdot e^{i2\pi(0.4)}, 0.1 \cdot e^{i2\pi(0.8)} \rangle \\ &\quad \oplus \langle 0.1 \cdot e^{i2\pi(0.6)}, 0.8 \cdot e^{i2\pi(0.6)} \rangle\} \\ &= \min\{\langle 0.3 \cdot e^{i2\pi(0.6)}, 0.5 \cdot e^{i2\pi(0.7)} \rangle, \langle 0.7 \cdot e^{i2\pi(0.64)}, 0.08 \cdot e^{i2\pi(0.48)} \rangle\} \end{aligned}$$

Using the AF of CFFN, we have AF $(\langle 0.3 \cdot e^{i2\pi(0.6)}, 0.5 \cdot e^{i2\pi(0.7)} \rangle) = 0.355$ and. AF $(\langle 0.7 \cdot e^{i2\pi(0.64)}, 0.08 \cdot e^{i2\pi(0.48)} \rangle)) = 0.358$. Clearly the minimum value occurs for $m = 1$.

$\therefore$ $d_3 = \langle 0.3 \cdot e^{i2\pi(0.6)}, 0.5 \cdot e^{i2\pi(0.7)} \rangle$ and v_3 is labelled as $[\langle 0.3 \cdot e^{i2\pi(0.6)}, 0.5 \cdot e^{i2\pi(0.7)} \rangle, 1]$.

Iteration 3

Now, the predecessor for node 4 is nodes 2 and 3. For the path 2–4, 3–4 ($m = 2, 3$ and $n = 4$), we have

$$\begin{aligned} d_4 &= \min\{d_2 \oplus d_{24}, d_3 \oplus d_{34}\} \\ &= \min\{\langle 0.7 \cdot e^{i2\pi(0.4)}, 0.1 \cdot e^{i2\pi(0.8)} \rangle \oplus \langle 0.2 \cdot e^{i2\pi(0.2)}, 0.9 \cdot e^{i2\pi(0.4)} \rangle, \\ &\quad \langle 0.3 \cdot e^{i2\pi(0.6)}, 0.5 \cdot e^{i2\pi(0.7)} \rangle \oplus \langle 0.3 \cdot e^{i2\pi(0.3)}, 0.8 \cdot e^{i2\pi(0.7)} \rangle\} \\ &= \min\{\langle 0.70 \cdot e^{i2\pi(0.41)}, 0.09 \cdot e^{i2\pi(0.32)} \rangle, \langle 0.37 \cdot e^{i2\pi(0.61)}, 0.4 \cdot e^{i2\pi(0.49)} \rangle\} \end{aligned}$$

AF $(\langle 0.70 \cdot e^{i2\pi(0.41)}, 0.09 \cdot e^{i2\pi(0.32)} \rangle) = 0.222$ and AF $(\langle 0.37 \cdot e^{i2\pi(0.61)}, 0.4 \cdot e^{i2\pi(0.49)} \rangle) = 0.229$.

Clearly the minimum value occurs for $m = 3$.

Therefore, $d_4 = \langle 0.70 \cdot e^{i2\pi(0.41)}, 0.09 \cdot e^{i2\pi(0.32)} \rangle$ and v_4 is labelled as $[\langle 0.70 \cdot e^{i2\pi(0.41)}, 0.09 \cdot e^{i2\pi(0.32)} \rangle, 3]$.

Iteration 4

Now, the predecessor for node 5 is nodes 2, 3, and 4. For the path 2–5, 3–5, 4–5 (m = 2, 3 and $n = 5$), we have

$$\begin{aligned} d_5 &= \min\{d_2 \oplus d_{25}, d_3 \oplus d_{35}, d_4 \oplus d_{45}\} \\ &= \min\{\langle 0.7 \cdot e^{i2\pi(0.4)}, , 0.1 \cdot e^{i2\pi(0.8)}\rangle \oplus \langle 0.4 \cdot e^{i2\pi(0.8)}, 0.8 \cdot e^{i2\pi(0.2)}\rangle, \\ &\langle 0.3 \cdot e^{i2\pi(0.6)}, 0.5 \cdot e^{i2\pi(0.7)}\rangle \\ &\oplus \langle 0.7 \cdot e^{i2\pi(0.5)}, 0.2 \cdot e^{i2\pi(0.7)}\rangle, \langle 0.70 \cdot e^{i2\pi(0.41)}, 0.09 \cdot e^{i2\pi(0.32)}\rangle \\ &\oplus \langle 0.2 \cdot e^{i2\pi(0.4)}, 0.6 \cdot e^{i2\pi(0.9)}\rangle\} \\ &= \min\left\{ \begin{array}{c} \langle 0.72 \cdot e^{i2\pi(0.81)}, 0.08 \cdot e^{i2\pi(0.16)}\rangle, \langle 0.71 \cdot e^{i2\pi(0.67)}, 0.1 \cdot e^{i2\pi(0.49)}\rangle, \\ \langle 0.70 \cdot e^{i2\pi(0.50)}, 0.05 \cdot e^{i2\pi(0.28)}\rangle \end{array} \right\} \end{aligned}$$

AF $(\langle 0.72 \cdot e^{i2\pi(0.81)}, 0.08 \cdot e^{i2\pi(0.16)}\rangle) = 0.454$; AF $(\langle 0.71 \cdot e^{i2\pi(0.67)}, 0.1 \cdot e^{i2\pi(0.49)}\rangle) = 0.388$ and AF $(\langle 0.70 \cdot e^{i2\pi(0.50)}, 0.05 \cdot e^{i2\pi(0.28)}\rangle) = 0.245$.Clearly the minimum value occurs for $m = 4$. Therefore, $d_5 = \langle 0.70 \cdot e^{i2\pi(0.50)}, 0.05 \cdot e^{i2\pi(0.28)}\rangle$ and ν_5 is labelled as $[\langle 0.70 \cdot e^{i2\pi(0.50)}, 0.05 \cdot e^{i2\pi(0.28)}\rangle, 4]$.

Iteration 5

Now, the predecessor for node 6 is nodes 4 and 5. For the path 4–6, 5–6 ($m = 4$, 5 and $n = 6$), we have

$$\begin{aligned} d_6 &= \min\{d_4 \oplus d_{46}, d_5 \oplus d_{56}\} \\ &= \min\{\langle 0.70 \cdot e^{i2\pi(0.41)}, 0.09 \cdot e^{i2\pi(0.32)}\rangle \oplus \langle 0.5 \cdot e^{i2\pi(0.6)}, 0.4 \cdot e^{i2\pi(0.5)}\rangle \\ &\langle 0.70 \cdot e^{i2\pi(0.50)}, 0.05 \cdot e^{i2\pi(0.28)}\rangle \oplus \langle 0.8 \cdot e^{i2\pi(0.6)}, 0.5 \cdot e^{i2\pi(0.3)}\rangle\} \\ &= \min\{\langle 0.75 \cdot e^{i2\pi(0.64)}, 0.03 \cdot e^{i2\pi(0.16)}\rangle, \langle 0.75 \cdot e^{i2\pi(0.67)}, 0.02 \cdot e^{i2\pi(0.14)}\rangle\} \end{aligned}$$

AF $(\langle 0.75 \cdot e^{i2\pi(0.64)}, 0.03 \cdot e^{i2\pi(0.16)}\rangle) = 0.261$ and AF $(\langle 0.75 \cdot e^{i2\pi(0.67)}, 0.02 \cdot e^{i2\pi(0.14)}\rangle) = 0.362$. Therefore, $d_6 = \langle 0.75 \cdot e^{i2\pi(0.67)}, 0.02 \cdot e^{i2\pi(0.14)}\rangle$ and ν_6 is labelled as $[\langle 0.75 \cdot e^{i2\pi(0.67)}, 0.02 \cdot e^{i2\pi(0.14)}\rangle, 5]$.

Since d_6 is the end node. So, the Fermatean fuzzy shortest route from ν_1 to ν_6 is $\langle 0.75 \cdot e^{i2\pi(0.67)}, 0.02 \cdot e^{i2\pi(0.14)}\rangle$. The following is how to determine the shortest route: ν_6 is labelled as $[\langle 0.75 \cdot e^{i2\pi(0.67)}, 0.02 \cdot e^{i2\pi(0.14)}\rangle, 5]$. ν_5 is labelled as $[\langle 0.70 \cdot e^{i2\pi(0.50)}, 0.05 \cdot e^{i2\pi(0.28)}\rangle, 4]$. ν_4 is labelled as $[\langle 0.70 \cdot e^{i2\pi(0.41)}, 0.09 \cdot e^{i2\pi(0.32)}\rangle, 2]$. ν_2 is labelled as $[\langle 0.7 \cdot e^{i2\pi(0.4)}, 0.1 \cdot e^{i2\pi(0.8)}\rangle, 1]$.

Hence, the shortest path is $1 \longrightarrow 2 \longrightarrow 4 \longrightarrow 5 \longrightarrow 6$ with the CFFN value of distance being $\langle 0.75 \cdot e^{i2\pi(0.67)}, 0.02 \cdot e^{i2\pi(0.14)}\rangle$.

In Fig. 5.7, the dotted line represents the shortest path and Table 5.15 provides the path of different nodes.

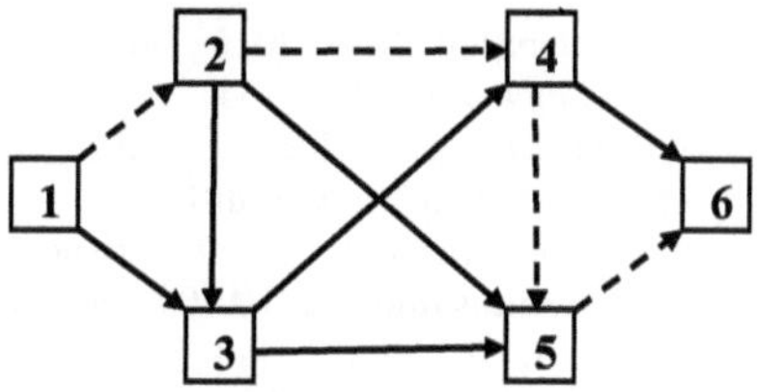

Fig. 5.7 CFF shortest path

Table 5.15 CFSP iterations

Nodes No. (N)	d_m	Shortest path from 1st node to nth node
2	$\langle 0.7 \cdot e^{i2\pi(0.4)}, 0.1 \cdot e^{i2\pi(0.8)} \rangle$	$1 \longrightarrow 2$
3	$\langle 0.3 \cdot e^{i2\pi(0.6)}, 0.5 \cdot e^{i2\pi(0.7)} \rangle$	$1 \longrightarrow 3$
4	$\langle 0.70 \cdot e^{i2\pi(0.41)}, 0.09 \cdot e^{i2\pi(0.32)} \rangle$	$1 \longrightarrow 2 \longrightarrow 4$
5	$\langle 0.70 \cdot e^{i2\pi(0.50)}, 0.05 \cdot e^{i2\pi(0.28)} \rangle$	$1 \longrightarrow 2 \longrightarrow 4 \longrightarrow 5$
6	$\langle 0.75 \cdot e^{i2\pi(0.67)}, 0.02 \cdot e^{i2\pi(0.14)} \rangle$	$1 \longrightarrow 2 \longrightarrow 4 \longrightarrow 5 \longrightarrow 6$

5.6 Conclusions

In this article, we proposed the innovative concept of CFFG as a generalization of CPyFG and its order, degree, and Cartesian product are discussed. We aim to apply our concept to a network optimization problem. We explored shortest path problem under complex environment and proposed a generalized Dijkstra's algorithm for solving CFFSPP based on the arithmetic operations and score function of CFFNs. The feasibility and viability of proposed algorithm are discussed using a numerical example. In future, we can extend the use of the proposed technique to more challenging network problems.

References

Akram M, Naz S (2019) A novel decision-making approach under complex Pythagorean fuzzy environment. Math Comput Appl 24:73 (CrossRef)

Akram M, Bashir A, Garg H (2020) Decision-making model Under complex picture fuzzy Hamacher aggregation operators. Comput Appl Math 39(3) (Springer Science and Business Media LLC). https://doi.org/10.1007/s40314-020-01251-2

Alkouri A, Salleh A (2012) Complex intuitionistic fuzzy sets. AIP Conf Proc 14:464–470

Atanassov KT (1986) Intuitionistic fuzzy sets. Fuzzy Sets Syst 20(1):87–96

Baba L (2013) Shortest path problem on intuitionistic fuzzy network. Ann Pure Appl Math 5(1):26–36

Buckley JJ (1989) Fuzzy complex numbers. Fuzzy Sets Syst 33:333–345

Chinnadurai V, Thayalan S, Bobin A (2021) Multi-criteria decision-making in complex Fermatean fuzzy environment. J Math Comput Sci (SCIK Publishing Corporation). https://doi.org/10.28919/jmcs/6495

Gani AN, Jabarulla MM (2010) On searching intuitionistic fuzzy shortest path in a network. Appl Math Sci 4(69):3447–3454

Garg H, Mahmood T, Rehman U, Ali Z (2021) CHFS: complex hesitant fuzzy sets-their applications to decision making with different and innovative distance measures. In CAAI transactions on intelligence technology, vol 6, issue 1. Institution of Engineering and Technology (IET), pp 93–122. https://doi.org/10.1049/cit2.12016

Greenfield S, Chiclana F, Dick S (2016) Interval-valued complex fuzzy logic. In: Fuzzy systems (FUZZ-IEEE), 2016 IEEE international conference on 2016

Jan N, Aslam M, Ullah K, Mahmood T, Wang J (2019) An approach towards decision making and shortest path problems using the concepts of interval-valued Pythagorean fuzzy information. Int J Intell Syst 34(10):2403–2428 (Wiley). https://doi.org/10.1002/int.22154

Khorsandi A, Liu X-C, Cao B-Y (2016) A new algorithm to shortest path problem with fuzzy arc lengths. In: International workshop on mathematics and decision science. Springer, pp 244–249

Luqman A, Akram M, Al-Kenani AN, Alcantud JCR (2019) A study on hypergraph representations of complex fuzzy information. Symmetry 11:1381. https://doi.org/10.3390/sym11111381

Ramot D, Milo R, Friedman M, Kandel A (2002) Complex fuzzy sets. IEEE Trans Fuzzy Syst 10:171186

Sahoo L (2021) Some score functions on Fermatean fuzzy sets and its application to bride selection based on TOPSIS method. Int J Fuzzy Syst Appl 10(3):18–29

Sanapati T, Yager YR (2019) Fermatean fuzzy sets. J Amb Intell Hum Comput 11:663–674

Senapati T, Yager RR (2019) Some new operations over Fermatean fuzzy numbers and application of Fermatean fuzzy WPM in multiple criteria decision making. Informatica 30(2):391–412

Shannon A, Atanassov K (1994) A first step to a theory of the intuitionistic fuzzy graphs. In: Lakov D (ed) Proceedings of the first workshop on fuzzy based expert systems. Sofia, pp 59–61

Shoaib M, Mahmood W, Albalawi W, Shami FA (2022) Notion of complex spherical fuzzy graph with application. In: Al-Olayan HA (ed) Journal of function spaces, vol 2022. Hindawi Limited, pp 1–27. https://doi.org/10.1155/2022/1795860

Thamizhendhi G, Kiruthica C, Suresh S (2011) Fermatean fuzzy hypergraph. J Hunan Univ Nat Sci 48(12):2333–2340

Thirunavukarasu P, Suresh R, Viswanathan KK (2016) Energy of a complex fuzzy graph. Int J Math Sci Eng Appl 10:243–248

Ullah K, Mahmood T, Ali Z, Jan N (2019) On some distance measures of complex Pythagorean fuzzy sets and their applications in pattern recognition. Compl Intell Syst

Yager RR (2014) Pythagorean membership grades in multicriteria decision making. IEEE Trans Fuzzy Syst 22:958–965

Yager RR, Abbasov AM (2013) Pythagorean membership grades, complex numbers and decision making. Int J Intell Syst 28(5):436–452

Yaqoob N, Akram M (2018) Complex neutrosophic graphs. Bull Comput Appl Math 6:85–109

Yaqoob N, Gulistan M, Kadry S, Wahab H (2019) Complex intuitionistic fuzzy graphs with application in cellular network provider companies. Mathematics 7:35 (CrossRef)

Yazdanbakhsh O, Dick S (2018) A systematic review of complex fuzzy sets and logic. Fuzzy Sets Syst 338:1–22 (Google Scholar) [CrossRef]

Zadeh LA (1965) Fuzzy sets. Inf Control 8(3):338–353

Zeng S, Shoaib M, Ali S, Abbas Q, Nadeem MS (2020) Complex vague graphs and their application in decision-making problems. In: IEEE access, vol 8. Institute of Electrical and Electronics Engineers (IEEE), pp 174094–174104. https://doi.org/10.1109/access.2020.3025974

Chapter 6
South African Perceptions on Artificial Intelligence in the Context of Morality and Sentience

Matthew Reid and Jean-Paul Van Belle

Abstract This study investigates South African people's perceptions around artificial intelligence (AI) in the context of morality and sentience. This was achieved by adapting and replicating a USA study conducted by the Sentience Institute in 2021 in South Africa. The adapted study contains 54 questions on AI, artificial sentience (AS) as well as general sentience. The results were analysed in three ways: descriptively, inferentially against demographic variables, and comparatively against the original study's results. Our study found that South African respondents have an underlying caution regarding AI/AS developments and see AI as an existential threat. Moreover, the respondents showed stronger compassion for the environment and non-human animals than for intelligent AI. Demographic factors such as age, gender, and religion play an influencing role in South African people's perceptions. Finally, South Africans appeared to have quite a different outlook to AI/AS developments compared to the USA population. This study underscores a need for further research into the nuances of these perceptions across different demographic groups and between the different cultural contexts.

Keywords Artificial sentience · AI morality perceptions · AI ethics · African attitudes towards AI

6.1 Introduction

The field of artificial intelligence (AI) has seen immense evolution, and this once deemed revolutionary technology has now transitioned to becoming a standard practice the norm: '*we now quickly tend to accept as normal was used to be seen as*

M. Reid · J.-P. Van Belle (✉)
Center for IT and National Development in Africa, University of Cape Town, Rondebosch, South Africa
e-mail: jean-paul.vanbelle@uct.ac.za

B. Shukla et al. (eds.), *Sustainable Technologies for Transforming Businesses and Societies*, Transactions on Computer Systems and Networks,
https://doi.org/10.1007/978-981-96-9471-6_6

extraordinary' (Haenlein and Kaplan 2019, p. 9). One of the possible next extraordinary steps includes the exploration of artificial sentience (AS) (Lavelle 2020). Although the general view is that no sentience or consciousness exists in current AI or robotics, there is still a need to research the possible ethics involved, such as the idea of 'robothood' (from the concept of personhood) which would involve legal rights and guidelines for potential future sentient robots or AI programs (Hildt 2019).

In the African context, there are initiatives to bring about awareness of AI readiness (Rapanyane and Sethole 2020), and strong growth around AI development can be seen in technology hubs and universities around Africa (Gwagwa et al. 2020). Although Africa is not at the forefront of AI development (Chilunjika et al. 2022), policy and further research is needed to ensure that Africa, and South Africa specifically, are ready for AI adoptions (Gwagwa et al. 2020; Schoeman et al. 2021). This research aims to address some of this gap.

The AIMS survey was launched in the USA to track perceptions around AI morality and sentiences over time. A descriptive analysis using South African responses therefore provides a substantive baseline. Inferential statistical analysis between USA and South Africa data also allows the exploration of possible difference cultural or contextual understandings and perceptions. This study hypothesizes that the data collected from the South African sample should differ from the data collected from the United States for the following reasons: (1) South Africa may have unique views on sentience, seen more as 'humanness', stemming from its philosophy of Ubuntu, which places more emphasis on the concept of humanity through connection. (Jecker et al. 2022); (2) South Africa likely has an overall negative perception of AI which comes from a fear that AI will replace the modern African workforce, threatening job security (Rapanyane and Sethole 2020).

6.2 Literature Review

6.2.1 Importance of Research into AI/AS/AM in the South African Context

Implementations of AI systems can result in positive economic growth, increased organizational efficiency, and workforce upliftment. This is no exception for South Africa and Africa as a whole (Schoeman et al. 2021). Global AI adoption is becoming increasingly imminent in South Africa (Segun 2021). To ensure that this AI expansion is conducted effectively, fairly, and ethically, policy and governance structures need to be in place (Haenlein and Kaplan 2019). Currently, AI policy is lacking in African countries (Gwagwa et al. 2020). In fact, most guidelines that are in place are inherently biased towards Euro-American perspectives (Segun 2021), but to ensure 'global' policies and guidelines are applicable and relevant to all areas where they apply, deliberate research efforts need to be focused in places such as Africa (Segun 2021). Africa has different unique philosophies and values, and a 'one size fits all' policy will struggle to be implemented effectively (Jecker et al. 2022). This is why African-specific research is relevant and necessary, and why active efforts need to be placed into hearing diverse opinions to ensure the global framework is in fact global (Segun 2021).

6.2.2 Artificial Sentience and Artificial Morality

The Sentience Institute defines 'artificial sentience' as the ability of an artificial entity to perceive both positive and negative experiences (Pauketat 2021). They do, however, admit to several shortcomings of this definition, such as the lack of scholarly use of the term, and that the term 'artificial' might detract from its ability to appear authentic.

Developing measurable artificial sentience systems is currently a great challenge in the AI landscape (Ashrafian 2015; Lavelle 2020), even though aspirations of creating a machine indistinguishable from man's intelligence date back to the 1950s (Haenlein and Kaplan 2019). Although there has been sizable research and development into AI, current AI is not seen to be sentient (Gibert and Martin 2021). Nevertheless, Chessman (2018) discusses the scenario where sentient AI receive similar legal protection to animals, having sentient AI classified as property. This will force the acceptance of responsibility upon the creator of the AI, thus enforcing cautious development. This will also limit potentially dangerous behaviour such as torture of sentient AI systems or heinous actions such as non-consensual molestation simulations, albeit as more a protection of the psychological effect that this behaviour will have on the person engaging with the AI system. Although Chessman does note that it is not a perfect application of legalization; it is a step in the correct direction in starting the conversation with policymakers around difficult topics such as AI protection for AI's sake and AI personhood.

Discussions around consciousness and sentience are not new in the AI space (Hildt 2019). Gibert and Martin (2021) raise the argument that sentience should be used as one of the convincing factors behind moral status, and that once something has sentience, it can more easily be granted a moral status. They continue to highlight that these discussions are difficult and often require inward retrospection, as sentience is more a theoretical than a practical discussion. Only once the definition of sentience has been decided as a whole, can other beings be measured against it. Once sentience can be attributed to other agents, then theorists can start defining more complex terms such as moral agency.

Morality is seen as a code of conduct put forward by a social group which defines the outline of acceptable behaviour for acting as an individual within the group (Gert and Gert 2022). Artificial morality (AM) more specifically refers to the ability of an AI system to exhibit moral capabilities (Misselhorn 2018) and suggests two criteria for artificial morality. Firstly, that a system can be classified as an agent, as in that it is self-originating in its actions and not affected by external factors. Secondly, that the AI system shows the capacity for acting with intent or reason. It should be noted, however, that human morality and artificial morality are inherently different due to the greater subjective opinions of the agent in question. Moreover, this limited definition of artificial morality does not equal the wider spectrum of full moral agency seen in humans. With the ever-increasing development of artificial intelligence, it is becoming increasingly important to define these concepts, as AI is getting to the stage where it will be making more and more decisions which will have to take these concepts into account (Misselhorn 2018; Shank and DeSanti 2018).

This relationship with AI and morality is incredibly complicated (Shank and DeSanti 2018), but this could be attributed to the complexities around morality as a field in its own right. Morality as a definition is simple, but the definitions of moral theories form a complex and ongoing discussion (Gert and Gert 2022). It is important to define these definitions as they will pave the way in how artificial intelligence is measured and studied, because much research is needed in understanding the relationship between AI and morality (Shank and DeSanti 2018).

As stated previously, artificial intelligence as we know it cannot be seen to be sentient (Gibert and Martin 2021). AI, as demonstrated in large language models and image processing, can be defined as 'narrow AI', since they are designed to do specific tasks and nothing more. Artificial general intelligence (AGI) is understood to be the broader understanding of intelligence, where the system is able to replicate the ability to learn new types of tasks to the level of that which would surpass typical human abilities (Ng and Leung 2020). While successes of narrow AI have been plentiful in recent years, the sentiment around AGI is slower, with little progress in the past few years (Bundy 2017). Taking advancements of AGI and sentience research into account, some scientists estimate that the probability that humanity will be able to create an AI system that is sentient is around 20% over the next 10 years (Chalmers 2023).

6.2.3 Current Perceptions of AI

There have been contradicting reactions to the modern implementation of AI technologies. The United States has banned certain types of facial recognition tools in public places. By contrast China has implemented these facial recognition AI tools to create a surveillance system which is tied to individual person's credit records and creates a demerit/reward system for 'good' behaviour among citizens. The European Union has responded to the increase in AI tools by increasing the regulation of the storage and processing of personal information (Haenlein and Kaplan 2019). The perceptions, whether positive or negative, seem to agree that AI with threaten business and society as we know it (Ergen 2019), and although it is agreed that AI will bring around economic growth, humanity will have to define the balance between privacy and this growth (Haenlein and Kaplan 2019).

Some recurring factors which have seen concern voiced include: the negative impact AI will have through the displacement of the workforce, ethical concerns around management of powerful AI tools, and the fear that humanity will lose control of AI, similar to scenarios seen in science fiction films (Fast and Horvitz 2016). Many of these factors have, however, been disproved or mitigated. Artificial intelligence theorists generally agree that AI will create more jobs than it displaces (Ergen 2019), and with all the developments that are taking place, equal risk mitigation strategies are being laid out to ensure humanity does not 'lose control' of AI (Mansell 2017).

Most studies provide a Western (i.e. Europe and North America) bias (Segun 2021). However, some international perceptions hold true: there is a prevalent negative perception in Africa around AI and the implementation of its technologies. The main concern is that AI will replace a large portion of the current workforce (Rapanyane and Sethole 2020). Positive sentiments are however echoed through strong growth in AI around African technology hubs, such as universities (Gwagwa et al. 2020), and studies agree that African AI expansion will bring around prosperity (Gwagwa et al. 2020; Rapanyane and Sethole 2020; Schoeman et al. 2021). Where African sentiments differ from international views are those around sentience and morality. Jecker et al. (2022) laid out that Africa has different views on sentience and morality compared to the traditional Western beliefs. They continue to explain that this unique sense of 'humanness' comes down to traditional values and concepts such as Ubuntu, which defines a philosophical outline for African 'humanity'.

6.2.4 The AIMS Study

However, there is a lack of research (especially African-focused research) into society's perceptions on AI morality, sentience, and legal rights. Pauketat et al. (2022) from the Sentience Institute looked to address the perceptions gap by grouping AI, Morality, and Sentience (AIMS) into a single survey. The AIMS survey is intended to measure these social perceptions in the United States (US) and periodically track changes over time. The survey included a comprehensive list of 94 questions grouped into six sections: Moral Consideration of AIs, Social Integration, Future Forecasts, Moral Consideration of Other Non-humans, Psychological Tendencies, and Demographics. The data collection period ran from November 2021 through to the end of December 2021. A total of 1232 responses were received, all from a balanced representative sample through sampling company Ipsos. Demographic information such as gender, age, region, ethnicity, income, and education were captured for analysis. Questions were based on a 7-point scale (from 'strongly disagree' to 'strongly agree') with a neutral 'no-opinion' mid-point.

In the study, most people agreed that sentient AI should be treated with respect and not intentionally harmed. The majority agreed that AI should be subservient to humans. Just under half of the population agreed that sentient AI should have moral agency. Interestingly, most people thought that AI is in some way threatening to the people living in the USA, as well as future generations. The study provided a comprehensive and substantive baseline sample for sentiments in the United States.

The AIMS study is relatively easy to replicate and allows for an in-depth comparative review between the US and SA, and it provides the baseline for South Africa. Hopefully it will be replicated in other country contexts as well, e.g. in India.

6.3 Research Methodology

This study used an empirical survey approach, with a mix of descriptive and inferential statistics. The descriptive approach was used when identifying key observations in the South African data and documenting these details (Bhattacherjee 2012). Inferential statistics were used to compare the South African results to those of the United States. The research instrument for this study was a survey created in Qualtrics that mimic a substantive part of the original AIMS survey on which this study is based. Table 6.1 summarizes the key constructs of the questionnaire. A Likert scale of 1–6 was used with 1 = agree completely and 6 = disagree completely except for techno-animism and substratism where the scale was reversed (as per US study) and the mind perception which was rated using a sliding scale from 0 to 100%.

Surveys were distributed to participants via an Instagram advertisement. Post data cleansing, 282 responses were recorded. The data collected was analysed independently, as well as comparatively, against the data in the original AIMS survey. The

Table 6.1 Constructs and scales used for questionnaire

Factor	Description	Questions	Scale
AS Caution	Caution towards artificial sentience (AS) developments	PMC1,3,4	Lower = agree
Pro-AS activism	Support for and willingness to advocate on behalf of AS	PMC2, 5–12	Lower = agree
AS treatment	Concern for the treatment of AS	MCE1-6	Lower = agree
Malevolence protection	Support for protecting AIs from malevolent actors and actions	MCE7-9	Lower = agree
Mind perception	Attribution of mind to current AIs	MP1-4	0–100% agree
Perceived threat	Perception of AIs as threatening	SI2-4	Lower = agree
Animal morality	Moral consideration of (non-human) living animals	MCA1-2	Lower = agree
Environmental morality	Moral consideration of the environment	MCE1-2	Lower = agree
Techno-animism	Belief that artificial entities can have a spiritual element	TA1-2	Higher = agree
Substratism	Prejudice against entities instantiated on non-carbon-based materials	SUB1-2	Higher = agree

following analysis/tests were run on the data: descriptive statistics, Cronbach alphas, factor analysis, analysis of variance (ANOVA) test, and t-tests.

The research was approved by the University's ethics committee. All participants received information regarding the research purpose. All participants gave consent to their data being used in this study, and all participant data aligned with current governance guidelines in terms of data protection, storage, and anonymity.

6.4 Data Analysis and Findings

The aim of this section is to dive into the data collected from the study. Firstly, the data cleansing process will be explained, a demographic analysis of the responses will be explored, and then the main results will be presented and organized by statistical test. After all non-valid responses were removed, a total of 282 valid responses remained.

6.4.1 Demographic Analysis

The majority of respondents indicated that they were between 18 and 30. There was a distinct gender bias in the responses with 178 identifying as female, 77 as male and 10 indicating 'Non-Binary', 'Not listed', 'Prefer not to say' or blank. With respect to religion, responses were grouped into larger categories. 'Christian—Protestant' and 'Christian—Catholic' were merged into a more general 'Christian' group. The three options 'No Religion', 'Atheist', and 'Agnostic' were grouped into 'Non-Religious'. Before grouping, the largest group was 'Christian—Protestant' with 64 responses. After grouping, Non-Religious' was the largest group with 103 responses. There were three blank responses recorded for religion. The political standing question was pulled from the original AIMS survey in the USA. Most respondents indicated that they had a very liberal (62) to liberal (64) political stance, with a strong moderate (55) view as well. Few identified themselves as being politically somewhat conservative (22) or very conservative (8), but this question had one of the highest instances of blank responses, with 71 blank responses recorded for political view.

6.4.2 Reliability and Validity Analysis

For reliability analysis, a Cronbach alpha test was run to check the internal consistency of the factors, and some constructs were removed to improve the reliability where applicable (Table 6.2).

An exploratory factor analysis with a fixed number of 11 factors (as per original research) was carried out to gauge how the questions correlated to their assigned factors. It was found that the questions and factors do in fact tie up to their assigned

Table 6.2 Cronbach alpha analysis

Factor	Original test items	Cronbach alpha (CA)	CA if least correlating item removed	Removed
AS caution	PMC 1, 3, 4	0.819	0.862	PMC1
Pro-AS activism	PMC 2, 5–12	0.891	0.900	PMC2
AS treatment	MCE 1–6	0.893	0.915	MCE6
Malevolence protection	MCE 7–9	0.870	0.928	MCE9
Mind perception (0–100%)	MP 1–4	0.735	Not better	
Perceived threat USA	SI 2, 3.1, 4	0.920	0.934	SI4
Perceived threat South Africa	SI 2, 3.2, 4	0.912	0.917	SI4
Moral consideration of non-human animals	MCA 1, 2	0.556		Only 2
Moral consideration of the environment	MCEn 1, 2	0.692		Only 2
Techno-animism (*)	TA 1, 2	0.583		Only 2
Substratism (*)	Sub 1, 2	0.492		Only 2

questions (table omitted due to space limitations). It also appeared that many of the factors could in fact be consolidated. 'Pro-AS Activism', 'AS Treatment', and 'Malevolence Protection' could mostly be combined excluding all the questions which fell outside factor one. Similarly, Moral Consideration of both Non-human Animal and the Environment could be combined. Conversely 'Pro-AS Activism' could be split into two smaller factors, namely PMC5–PMC6 and PMC7–PMC12 (with PMC2 merging into AS Caution).

An exploratory factor analysis was also performed, setting the cut-off for the number of factors to an eigenvalue of one. This grouped the questions into eight factors. If the study were to be replicated further, a recommendation could be to reanalyse the dataset using these eight factors.

6.4.3 Descriptive Analysis

When it comes to questions regarding caution around artificially sentient developments (AS caution), most of the respondents indicated that they were in support of global bans (mean = 3.0712; $n = 281$). This aligns with responses around the perceived threat around AS developments. Most respondents felt (agreed) that AS beings may be harmful to either them, the populations of SA or the USA, or future generations (Table 6.3).

Respondents were evenly split in terms of showing artificial sentience activism support. Certain questions indicated more support, such as PMC7 (protecting AS being in terms of scientific research—with a mean of 2.88), where majority veered towards disagreement. Some strong negative sentiments were seen, for example PMC12 (mean 4.36), around considering joining public demonstrations against the mistreatment of AS beings.

Regarding the treatment of artificially sentient beings, respondents again showed neither strong positive nor negative perceptions when aggregated. Some questions indicated agreement, such as MCE1 (deserving treatment of respect) and MCE5 (torture is wrong), whereas MCE6 showed disagreement (welfare of AI beings is one of the most important social issues). However, respondents indicated agreement when it comes to questions regarding the protection of AS beings from malevolent actors, especially from humans. All three questions indicated between agreement and somewhat agreement.

By comparison, most respondents agree that animals and the environment deserve to be included in the moral circle and that animal and environmental welfare is one of the most important social issues we face in the world today.

In terms of attributing 'mind' to AI, i.e. mind perception, respondents voted lower towards typically emotive characteristics, as seen in MP1 (emotions) and MP2 (feelings) compared to the more typically analytical characteristics, as seen in MP3 (analytical thinking) and MP4 (rationality). Note that this was measured on a sliding percentage attribution scale of 0–100%. A mean of 40.28% was calculated for the factor 'Mind Perception'.

For Techno-Animism, the interaction between humans and technology, respondents generally disagreed with the statements that artificial beings contain a spirit or that there is interaction between spirits of natural and AI beings. Lastly, respondents hovered around neutrality to the questions around substratism (or the discrimination

Table 6.3 Descriptive statistics for constructs

Factor (as per original survey)	No. of questions	Mean	Std. dev
AS caution	3	3.071	1.482
Pro-AS activism	9	3.608	1.244
AS Treatment	6	3.642	1.275
Malevolence protection	3	2.507	1.236
Perceived threat	3	2.606	1.314
Moral consideration of non-human animals	2	1.954	0.948
Moral consideration of the environment	2	1.525	0.708
Techno-animism (*)	2	2.270	1.476
Substratism (*)	2	4.130	1.700
Mind Perception (0–100%)	4	40.28%	21.36%

of sentient beings based on the substrate that shaped their sentient ability). They indicated slight agreement that AI beings will always count less than humans, but slight disagreement that humans have the right to use AI beings however they please.

6.4.4 *Influence of Demographics on Participant Views*

This section sets out to explore the influence of an individual's demographics on the dataset and to try to understand if these demographic factors are statically significant in understanding the differences in responses. In cases where there were only two groups being tested (such as gender, where only the two largest responses were considered) two tests were conducted. Firstly, a Levene's Test for Equality of Variances was carried out to test if the variances between the groups were approximately equal. Then a T-test was conducted to measure the significance between the two groups. When the demographic responses presented more than two significant options, three tests were conducted. Firstly, a test of Homogeneity of Variances based on the median value was carried out to ensure that the variances of each of the groups are approximately, and then an analysis of variance (ANOVA) test was used to gauge if the means across the groups were statistically significant. Lastly, a Tukey test was used to judge the statistical significance between the demographic groups. The following discusses only questions or factors which were statistically significant.

Age. Age was grouped into respondents who indicated that they were between 18 and 25 years old and those that indicated greater than 25 years of age. Mind Perception and Substratism were factors that presented statistical significance. For Mind Perception, the 18–25-year-olds showed higher levels of agreement to the statements that robots/AI have the ability to experience emotions/feelings/analytical reasoning/rationality compared to the greater than 25-year-olds. It is to be noted though that both agreed less than 50% to those statements (Table 6.4).

Gender. For gender analysis, the two largest response groups were analysed against each other, namely male and female. The factors which showed statistical significance were: AS Caution, AS Treatment, Malevolence Protection, and Techno-Animism. In terms of AS Caution, the male respondents indicated disagreement as opposed to female respondents who rather indicated more agreement towards 'caution towards artificial sentience (AS) developments'. For AS Treatment, both groups responded

Table 6.4 Age-factor-based *T*-test analysis and group descriptive statistics

Factor	Group	*N*	Mean	Std. dev	Std. error mean
Mind perception	18–25	145	42.84%	21.15%	1.756%
	25 +	119	36.57%	21.67%	1.986%
Substratism	18–25	147	4.340	1.584	0.131
	25 +	118	3.894	1.830	0.168

in slight disagreement to questions around 'concern for the treatment of AS beings' although males indicated slightly stronger sentiments towards more disagreement. Both groups agreed towards the 'support for protecting AIs from malevolent actors and actions' questions, with females showing more agreement compared to males who indicated only slight agreement. Lastly, for Techno-Animism, respondents both disagreed in the 'belief that artificial entities can have spirits', but males disagreed more compared to females (Table 6.5).

Religion. For analysis of the religion variable, the protestant and catholic group were combined into a larger Christian group, and the irreligious groups were placed into one non-religious group. Table 6.6 presents statistically significant ANOVA results. Only Moral Consideration of Non-human Animals and Substratism were statistically significant. In terms of Moral Consideration of Non-human Animals, both the Christian and non-religious group agreed with the 'consideration of non-human animals' questions. The non-religious group did indicate that they agreed more than the Christian group when rounding isn't considered. For Substratism, the Christian group showed statistical significance to both the non-religious and Buddhist groups. The Christian group showed more agreement to the 'prejudice against entities instantiated on non-carbon-based materials' questions, whereas the non-religious group teetered around neutrality, and the Buddhist group disagreed with the statements.

Table 6.5 Gender-factor-based *T*-test analysis and group descriptive statistics

Factor	Group	*N*	Mean	Std. dev	Std. error mean
AS caution	Male	77	3.649	1.602	0.183
	Female	177	2.864	1.358	0.102
AS Treatment	Male	77	3.909	1.357	0.155
	Female	178	3.531	1.228	0.092
Malevolence protection	Male	75	2.758	1.364	0.158
	Female	176	2.399	1.174	0.089
Techno-animism	Male	74	1.993	1.430	0.166
	Female	177	2.438	1.498	0.113

Table 6.6 Religion test analysis and group descriptive statistics

Factor	Group	*N*	Mean	Std. dev	Std. error
MC non-human animals	Christian	90	2.300	1.054	0.111
	Non-religious	103	1.791	0.862	0.085
Substratism	Buddhist	6	2.417	1.158	0.473
	Christian	89	4.714	1.662	0.176
	Non-religious	103	3.806	1.663	0.164

6.4.5 Comparison Between USA and SA

A T-test was done to test the hypothesis that South Africans have a different view on sentience compared to the population of the United States of America. The analysis was done on both the factor level and per individual question; only the former is reported on here. Eight of a possible ten factors were statistically significantly different when comparing the SA group against the USA group. AS Caution, Malevolence Protection, Mind Perception, Perceived Threat USA, both Moral Considerations, Techno-Animism, and Substratism all presented statistically significant results. Below is a summary of results.

AS Caution—In terms of 'caution towards artificial sentience (AS) developments', the USA group disagreed more than the SA group, who in fact showed agreement.

Malevolence Protection—For 'support for protecting AIs from malevolent actors and actions' the USA group again had opposing views to the SA group, indicating disagreement compared to the SA group's agreement.

Mind Perception—Although both groups indicated less than 50% for 'attribution of mind to current AIs', the USA group attributed a slight 4% higher with 44% as opposed to the SA group's 40%

Perceived Threat USA—In terms of the 'perception of AIs as threatening (US)', the two groups found themselves on opposing sides, with the USA group disagreeing to the idea of a threat, compared to the SA group's agreement.

Moral Considerations of Non-human Animals and the Environment—Similarly, the USA group disagreed to both factors, whereas the SA group agreed to both.

Techno-Animism—Although both groups disagreed with the statements in the 'belief that artificial entities can have spirits', the USA disagreed less than the SA group.

Substratism—Lastly, both groups were neutral in the 'prejudice against entities instantiated on non-carbon-based materials' statement, but the USA group agreed slightly and consistently more than the SA group.

6.5 Discussion/Implications

6.5.1 Interpretation of Findings

Exploring and understanding the interrelation of the descriptive statistics in the context of the survey highlighted the following key takeaways. South African people seem to have an underlying caution or negative perception around artificial intelligence as seen through factors 'AS Caution' and 'Perceived Threat SA', implying that they think that AI is an immediate danger or threat to them. This becomes interesting when considering the factors 'AS Treatment' and 'Malevolence Protection', because even though South African people seem to see AI as a threat, they don't have negative views on the treatment of AI, and in fact, they have positive sentiments around the protection of AI beings from malevolent actors.

The survey also highlighted the difference of opinion when it comes to the mind attribution of AI beings. South African people were much more likely to attribute characteristics such as 'thinking' or 'rationality' to AI beings as opposed to the likes of 'feeling' or 'emotions'. This difference of attribution shows that South African people have a more technical view of AI as opposed to thinking of it as a potentially sentient being. Lastly, the survey pointed out the difference in values of the South African population. They were much quicker to include the environment and non-human animals into the moral circle compared to AI beings.

6.5.2 Demographic Influence

The demographic variable T-tests and ANOVA tests indicated the prevalent effect on which these different demographic factors influence the worldview, perception, and understanding of South African people. Note that level of education or province did not influence views.

Looking at the effect of **age**, the study indicated the younger population had a higher mind attribution to AI beings compared to the older population. The younger group also seemed to have more consideration for the environment.

The influence of **gender** was also highlighted through the following observations. Female respondents opted for more caution over males when it comes to the development of AI. Male respondents, on the other hand, agreed more that AI beings should be subservient to humans, compared to the female respondents. This difference in perceptions begs the question 'what is it about gender which causes these differences?' Further research could dive into potential topics such as: the impact of the different societal roles males and females traditionally play in South Africa, and their corresponding impact on perceptions.

Lastly, **religion** also played a significant role in South African people's perceptions around AI. The following three differences showed statistical significance, each driven by different major South African religions: Christian respondents had much stronger prejudice (as seen through substratism) against non-carbon sentient beings, compared to the non-religious or Buddhist groups. Hindu respondents had much higher levels of attribution of characteristics like emotions/feeling to AI beings over most other religious groups.

6.5.3 Comparison to United States Study

There were stark differences in the opinions of the United States population versus the South African population. Of the ten relevant factors, eight showed statistical significance. Starting with perceptions around AS caution, the USA group disagreed that there should be future bans on AS developments, where the SA group agreed, showing more caution. Conversely, looking into to 'Malevolence Protection', the SA group showed support in protecting AS beings from malevolent actors, where the USA group disagreed that it was necessary. Similarly, with 'Perceived Threat USA/SA', the USA population disagreed that AI poses an immediate threat, where the SA population agreed and see AI as a threat. These opposing but mirroring views shed some insight into how the different populations perceive AI beings, but further investigation to understand the causal factors behind these results is indicated. In respect of general (i.e. non-artificial) sentience, the USA group also showed much less moral consideration to the environment and non-human animals, compared to the SA group. This begs the investigation into the broader question about how these different populations understand sentience and the natural world.

6.5.4 Hypothesis Testing

Tying the study base to the initial research hypotheses, what does the data imply about the impact of Ubuntu on perception as well as the recorded overall negative perceptions of AI in literature?

Ubuntu emphasizes the unique sense of humanness through connectedness. Jecker et al (2022) indicate that it is this philosophy of Ubuntu which affects South African people's view on humanness. To analyse this, 'Techno-animism' and 'Mind Perception' can be explored deeper. It can be seen through the results that South African people do not perceive AI beings as having a spirit and don't see them as capable of traditionally human characteristics such as emotions or feelings. This confirms hypothesis one by implying that South African people have a distinct view of separation between potentially sentient AI beings and humans. This does not however explicitly link this separation to the philosophy of Ubuntu, which could be explored in further studies.

Looking less at the philosophical level and more on to a pure perceptive level, the study also confirmed hypothesis two, that South African people have an overall negative perception of AI, especially when it comes to economic security (Jecker et al. 2022). This hypothesis can be confirmed when looking at factors such as 'AS Caution' and 'Perceived Threat SA'. South African people indicated caution when it comes to future AI developments, and they also see AI as a direct threat to themselves, the future generations, and South Africa as a whole. This fear warrants further research into why the respondents felt that way; similar to hypothesis one, further research needs to be explored to understand why.

6.6 Conclusion

This study undertook the broad task of understanding South African people's perceptions of artificial intelligence (AI) in terms of morality and artificial sentience (AS). Using a quantitative approach, it replicated an existing United States of America based study presented by the Sentience Institute and explored a wide range of questions from caution in AS developments to substratism.

The findings revealed an underlying negative sentiment towards AI beings by South African people. This was seen through factors 'AS Caution', which advocates for caution towards future AI/AS development and 'Perceived Threat SA', which implies that South African people see AI as an immediate threat to themselves, their future generations, and South Africa as a whole. Despite these negative sentiments, South African people still presented a significant support to protect sentient AI beings from malevolent actors. This duality highlights the complex relationship South African people have with the theories of AI or sentience in general.

Next, the study compared the perceptions of the South African respondents with the original AIMS United States respondents. Several factors showed contrasting views on artificial sentience as well as general sentience between these two populations. While the USA population didn't see a need to develop AI with caution and didn't see it as an immediate threat, the SA population clearly indicated a need for caution and does see AI as an immediate threat. However, the SA group showed support for protecting AS beings from malevolent actors, whereas the USA group did not. Lastly, in terms of general sentience, the USA population showed much less moral consideration to the environment and non-human animals compared to the South African group.

This study also looked at the impact of individual demographic factors on people's perceptions. Factors such as age, gender, and religion all showed statistical influence on several of South African people's perceptions around AI/AS developments.

Lastly, the original hypotheses were validated. South African people do have a unique understanding of sentience; whether or not this can be attributed to the concept of Ubuntu warrants further research. Similarly, South African people do have an underlying negative perception of AI, but again further research will have to be carried out to understand where this fear stems from.

The following **limitations** were observed. The demographic analysis indicated that the sample is not necessarily truly representative of South Africa; e.g. some provinces had no respondents. The issue of survey bias is also a concern, due to the fact that these are all self-reported responses.

Future research could validate, replicate, and expand the current research. Now that a baseline is set in South Africa, this study could be repeated over a period of a few years to combine all the data into a longitudinal study. Given that AI is becoming more prevalent and more advanced, it will be interesting to track perceptions over time. Secondly, after confirming that demographics affect people's perceptions, future studies could more fully explore the correlation or causation effects of demographic variables. Alternatively, qualitative questions could follow specific quantitative questions to explore and quantify the differences. This could help understand the ‘why’ behind some of these questions rather than just the ‘what’. This applies particular to understanding why South Africans fear future AI more, as well as their stronger views on the differences between animal/human and technology-based sentience. Finally, residents in countries with different economic, political, cultural, and other contexts are likely to hold views on AI morality and sentience that are different from South Africa or the USA. Thus, a replication study in the Indian context could yield quite interesting findings.

In **conclusion**, the research has shed some light on the complex relationship between South African people and their perceptions around artificial intelligence. While there are underlying characteristics of caution and fear in South African people’s perceptions, there is also a prevalent sense of empathy towards sentient beings—artificial or not. These conflicting emotions could be attributed to demographic factors such as age, gender, and religion, or even cultural philosophies such as Ubuntu. While this study did confirm the hypotheses, it emphasized a definite need for further research and understanding into these fields. Artificial intelligence is becoming more and more integral in our daily lives, and understanding these nuances will be crucial for building the frameworks of the future.

References

Ashrafian H (2015) Artificial intelligence and robot responsibilities: innovating beyond rights. Sci Eng Ethics 21(2):317–326. https://doi.org/10.1007/s11948-014-9541-0

Bhattacherjee A (2012) Social science research: principles, methods, and practices, 2nd edn. CreateSpace

Bundy A (2017) Preparing for the future of artificial intelligence. AI Soc 32(2):285–287. https://doi.org/10.1007/s00146-016-0685-0

Chalmers DJ (2023) Could a large language model be conscious? arXiv. https://doi.org/10.48550/arxiv.2303.07103. Accessed 21 Nov 2024

Chessman CF (2018) Not quite human: artificial intelligence, animals, and the regulation of sentient property. Soc Sci Res Netw. https://doi.org/10.2139/ssrn.3200802

Chilunjika A, Intauno K, Muzvidziwa-Chilunjika SRT (2022) Artificial intelligence and public sector human resource management in South Africa: opportunities, challenges and prospects. SA J Human Resour Manag 20. https://doi.org/10.4102/sajhrm.v20i0.1972

Ergen M (2019) What is artificial intelligence? Technical considerations and future perception. Anatol J Cardiol 22. https://doi.org/10.14744/anatoljcardiol.2019.79091

Fast E, Horvitz E (2016) Long-term trends in the public perception of artificial intelligence. Natl Conf Artif Intell 31(1):963–969. https://ojs.aaai.org/index.php/AAAI/article/view/10635/10494

Gert B, Gert J (2022) The definition of morality. In Zalta EN (ed) The Stanford encyclopedia of philosophy. https://plato.stanford.edu/entries/morality-definition/

Gibert M, Martin DA (2021) In search of the moral status of AI: why sentience is a strong argument. AI & Soc 37(1):319–330. https://doi.org/10.1007/s00146-021-01179-z

Gwagwa A, Kraemer-Mbula E, Rizk N, Rutenberg I, De Beer J (2020) Artificial intelligence (AI) deployments in Africa: benefits, challenges and policy dimensions. Afr J Inform Commun 26. https://doi.org/10.23962/10539/30361

Haenlein M, Kaplan AM (2019) A brief history of artificial intelligence: on the past, present, and future of artificial intelligence. Calif Manage Rev 61(4):5–14. https://doi.org/10.1177/0008125619864925

Hildt E (2019) Artificial intelligence: does consciousness matter? Front Psychol 10. https://doi.org/10.3389/fpsyg.2019.01535

Jecker NS, Atuire C, Ajei MO (2022) The moral standing of social robots: untapped insights from Africa. Philos Technol 35(2). https://doi.org/10.1007/s13347-022-00531-5

Lavelle S (2020) The machine with a human face: from artificial intelligence to artificial sentience. Adv Inform Syst Eng Workshops 63–75. https://doi.org/10.1007/978-3-030-49165-9_6

Mansell R (2017) Are we losing control? InterMedia 45(3):4–7. http://eprints.lse.ac.uk/85640/

Misselhorn C (2018) Artificial morality. concepts, issues and challenges. Society 55(2):161–169. https://doi.org/10.1007/s12115-018-0229-y

Ng GW, Leung WC (2020) Strong artificial intelligence and consciousness. J Artif Intell Conscious 07(01):63–72. https://doi.org/10.1142/s2705078520300042

Pauketat JVT, Ladak A, Anthis JR (2022) Artificial intelligence, morality, and sentience (AIMS) survey: 2021. Sentience Institute. https://doi.org/10.31234/osf.io/dzgsb

Pauketat JVT (2021) The terminology of artificial sentience. Sentience Institute (2021). https://doi.org/10.31234/osf.io/sujwf

Rapanyane MB, Sethole FR (2020) The rise of artificial intelligence and robots in the 4th Industrial Revolution: implications for future South African job creation. Contemp Social Sci 15(4):489–501. https://doi.org/10.1080/21582041.2020.1806346

Schoeman W, Moore R, Seedat Y, Chen JY (2021) Artificial intelligence: is South Africa ready? [White paper]. In: Working papers. Gordon Institute of Business Science, Accenture. https://www.gibs.co.za/about-us/faculty/documents/whitepapers/chen_ai%20is%20south%20africa%20ready.pdf. Accessed 21 Nov 2024

Segun ST (2021) Critically engaging the ethics of AI for a global audience. Ethics Inf Technol 23(2):99–105. https://doi.org/10.1007/s10676-020-09570-y

Shank DB, DeSanti A (2018) Attributions of morality and mind to artificial intelligence after real-world moral violations. Comput Hum Behav 86:401–411. https://doi.org/10.1016/j.chb.2018.05.014

MIX
Papier aus verantwortungsvollen Quellen
Paper from responsible sources
FSC® C105338

If you have any concerns about our products,
you can contact us on
ProductSafety@springernature.com

In case Publisher is established outside the EU,
the EU authorized representative is:
Springer Nature Customer Service Center GmbH
Europaplatz 3, 69115 Heidelberg, Germany

Printed by Libri Plureos GmbH
in Hamburg, Germany